Math for Muggles

Luis Tejada

Table of Contents

Prologue

This book has been created with a clear and simple purpose: to make mathematics accessible to everyone, regardless of their previous experience with the subject. Here, we will dismantle the idea that mathematics is an impenetrable mystery and show that it is a powerful and surprisingly friendly tool.

Mathematics is much more than numbers, equations, and complicated formulas. It is the universal language of logic, problem-solving, and creativity. Often, people feel intimidated by math because of a negative past experience or simply because it was never explained to them properly. This book aims to change that.

Throughout these pages, we will explore mathematical concepts in a clear and accessible way. From whole numbers to geometry, from fractions to probability, we have distilled these ideas into their simplest form. Our mission is for everyone to be able to understand and enjoy mathematics.

Make no mistake—we're not saying that math is easy, but we firmly believe that it can be understood by everyone. As you move forward on this journey, you will discover that mathematics is a powerful tool that can help you solve everyday problems, make informed decisions, and appreciate the beauty of the patterns and relationships that surround us.

This book is for those who have ever felt that math was out of their reach, for those who never felt comfortable with numbers, and for those who want to rediscover the wonder found in the marvels of mathematics.

1.Integers: Positive and negative numbers that you can count   on a number line, like -3, -2, -1, 0, 1, 2, 3, etc.

Mathematics is a universal language that allows us to describe and understand the world around us. On this fascinating mathematical journey, we encounter a fundamental tool: integers. These positive and negative numbers form a kind of "mathematical lifeline," allowing us to represent a wide range of real-life situations.

Imagine this line of integers as a long path stretching infinitely in both directions. Along this path, you find whole numbers—numbers that are not divided into smaller parts—like episodes of a series moving forward and backward, weaving a rich and varied mathematical narrative.

On this line, you can find numbers such as -3, -2, -1, 0, 1, 2, 3, and many more. Zero is the reference point, the epicenter of this line, from which you can explore both the positive and negative directions. If you journey to the right from 3, you'll encounter 2, then 1, and eventually arrive at 0. But the line doesn't end there—it continues with negative numbers, like -1 and -2.

If you venture to the left from 3, you'll discover a series of negative numbers, such as -2 and -1, before returning to zero. This "lifeline" of integers is like a river that flows in both directions, allowing us to capture the duality of the mathematical world.

Integers are versatile and essential. We use them to describe a variety of everyday situations. They can represent gains and losses, hot and cold temperatures, positions forward and backward in time, debts and savings, and many other things. Moreover, they serve as the foundation upon which more advanced mathematical concepts are built. Integers are fundamental and versatile, used to describe a wide range of scenarios—from financial transactions and weather patterns to historical timelines and personal finances.

They are also the basis for more advanced mathematical ideas, such as rational numbers, real numbers, and concepts from algebra, arithmetic, and geometry. Integers are essential for understanding and performing more complex mathematical operations and are used across numerous disciplines, including physics, economics, computer science, and statistics.

Integers play a crucial role in mathematics and are vital tools for understanding and solving problems both in daily life and in academic and scientific fields.

So, the next time you encounter integers in your mathematical adventures, remember that they are far more than just simple symbols on a page. They are the line that connects the bright and dark moments of mathematics, and they are here to guide and accompany you every step of the way.

What Are Integers?

Integers are a set of numbers that include both positive and negative numbers, along with zero. Integers can be represented on a number line where zero sits in the center, positive numbers extend to the right, and negative numbers extend to the left. This number line is commonly used to visually represent integers.

Positive Integers:

These are all integers greater than zero. On a common number line, they are located to the right of zero. Examples of positive integers include 1, 2, 3, and so   on.
While we often visualize them to the right of zero on a number line, mathematically speaking, positive integers are simply those greater than zero, regardless of their position in any diagram.

Zero:
Zero is neither positive nor negative and is represented as "0."
On the integer number line, zero is the unique starting point. From zero, you can count to the right to reach positive numbers (1, 2, 3, ...) and to the left to   reach   negative   numbers   (-1,   -2,   -3,   ...).
In summary, zero is a neutral integer, placed at the center of the number line, acting as the dividing point between positives and negatives.

Negative Integers:

These are all integers less than zero and are located to the left of zero on the number   line.   Examples   include   -1,   -2,   -3,   and   so   on.
Negative integers represent quantities less than zero and are used in mathematics and other fields to denote debts, sub-zero temperatures, and other situations where the value is less than nothing.

In short, integers include all positive numbers, negative numbers, and zero. They are used in mathematics and many other disciplines to represent whole quantities and can be operated on using addition, subtraction, multiplication, division, and other mathematical operations.

2.**Addition**: The operation of combining numbers to get a total

Addition is a fundamental mathematical operation used to combine numbers or quantities to obtain a total or accumulated amount. When you add two or more numbers, you are putting them together. For example, if you add the numbers 3 and 4, you get a total of 7:

$3 + 4 = 7$

Addition is one of the four basic arithmetic operations, along with subtraction, multiplication, and division. It is used in a wide variety of contexts, from simple everyday calculations to more complex applications in fields such as science, engineering, economics, and statistics.

Addition is one of the four basic arithmetic operations, and its importance lies in its versatility and applicability across a broad range of contexts. Here is an expanded explanation of its relevance:

Everyday Life

Addition is used in everyday situations to calculate prices, add quantities in a shopping list, determine the total amount spent in a store, calculate a restaurant bill, or add up hours worked at a job. In short, it is an essential skill for financial management and daily activities.

In daily life, addition is a key skill for financial management and completing numerous tasks. Here are some more examples of how addition is applied in everyday situations:

Personal Budgeting: Addition is used to track personal income and expenses. It helps people calculate how much money they have left after paying monthly bills.

Shopping and Expenses: When making purchases, addition is used to calculate the total cost of items or services, including taxes and discounts. It's also used to determine how much has been spent over a month.

Splitting Expenses: When sharing a bill with friends or roommates, addition helps to divide the cost fairly among everyone.

Inventory Control: In businesses and homes, addition is used to keep track of inventory, calculating how many items are available and how many have been sold or consumed.

Calculating Change: When paying with cash, addition helps determine the change to be received after a purchase.

Event Planning: While organizing events like parties or meetings, addition is used to calculate total costs, including food, drinks, decorations, and other expenses.

Time Management: In situations that require tracking time, such as scheduling tasks or recording work hours, addition is used to calculate the total time spent on an activity.

Addition is an essential skill applied in countless daily situations, whether for financial control, shopping, event planning, or time management. It is a fundamental tool for making informed decisions and efficiently performing everyday tasks.

School Mathematics

Addition is one of the first mathematical operations students learn and is fundamental in elementary arithmetic. It helps students develop number skills, understand number structures, and learn concepts like commutativity and associativity.

It is one of the most fundamental mathematical operations and among the first learned in school. Its importance in school mathematics is significant for several reasons:

Foundation of Arithmetic: Addition is a key base of arithmetic, the branch of math focused on number operations. Learning to add correctly is critical for developing more advanced mathematical skills.

Development of Number Skills: Addition helps students build basic number skills, such as recognizing numbers and numerical patterns.

Number Building: Students learn how numbers are built through addition. For example, they understand that 7 can be made by adding 3 and 4 or 2 and 5. This is essential for understanding the structure of numbers.

Mathematical Properties: Working with addition also teaches important properties, such as commutativity (order doesn't affect the result) and associativity (grouping doesn't affect the result). These apply to more advanced math operations.

Preparation for Later Operations: Addition lays the groundwork for subtraction, multiplication, and division. Understanding addition is essential for mastering these other operations.

Problem Solving: Addition is used in solving math problems and real-world problems involving quantities and totals.

In summary, addition is a foundational operation in school math that not only enables students to perform calculations but also contributes to broader mathematical development, such as logical thinking and number sense. This strong foundation is essential for success in math as students progress in their education.

Sciences

In scientific disciplines such as physics and chemistry, addition is used to calculate quantities such as speed, force, and energy. In statistics, addition is essential for computing averages, standard deviations, and other important measures.

Addition plays a crucial role across various scientific disciplines and in statistics:

Physics: In physics, addition is used to calculate and combine physical quantities. For example, to calculate an object's speed, you add distances covered over time intervals. In acceleration, applied forces may be summed or integrated. Addition is also fundamental in mechanics, where it is used to describe the motion of objects under various forces.

Chemistry: In chemistry, addition is used in stoichiometry, which deals with the quantitative relationships of reactants and products in chemical reactions. The stoichiometric coefficients of substances in a chemical equation are added to balance the reaction, ensuring the law of mass conservation is upheld.

Statistics: Addition is foundational in statistics. It's used to compute descriptive statistics such as the mean (average), which involves summing all values and dividing by the number of observations. It is also used for calculating sum of squares, sum of products, and other values crucial for computing standard deviation and linear regression.

Biology and Health Sciences: In biomedical research, addition is used to combine data from multiple experiments or measurements. For instance, in clinical trials, participant results are added to produce statistics on treatment effectiveness.

Geology and Environmental Science: In these fields, addition is used to calculate values such as accumulated rainfall, soil erosion, or average temperatures over time. These calculations are vital for studying and managing natural resources.

Addition is a powerful mathematical tool applied in a wide range of scientific contexts to perform calculations, analyze data, and understand quantitative relationships. Its use in these disciplines underscores its importance as a mathematical foundation in scientific research and advancing knowledge of the natural world.

Engineering

In engineering, addition is used in the design and analysis of structures, electrical circuits, mechanical systems, and more. It is also employed in performance evaluations of complex systems and processes.

Addition is an essential tool in engineering applications and is used in the design, analysis, and evaluation of a wide variety of systems and components:

Structural Design: Engineers use addition to calculate and sum forces, stresses, and loads in structures like bridges, buildings, and other constructions. This ensures structures are safe and can support expected loads.

Electrical Circuit Analysis: In electrical engineering, addition is used to calculate currents and voltages in circuits, which is essential for designing efficient and safe electrical systems.

Mechanical Design: Addition is applied in analyzing mechanical systems, including machines, motors, and transmissions. Equations involving addition are used to calculate moments, forces, and velocities of components.

Fluid Dynamics: In fluid engineering, addition helps evaluate pressures, flows, and speeds in systems like pipes, pumps, and valves. It is also applied in aerodynamics design in the aerospace industry.

Process Control and Automation: In control engineering, addition is used to design systems that regulate variables such as temperature, pressure, and flow in industrial processes.

Performance Evaluation: Engineers use addition in performance calculations to evaluate the efficiency and behavior of complex systems—key for optimizing industrial processes and systems.

Simulation and Numerical Modeling: Addition is integral in numerical models and computational simulations in engineering. It's used to discretize differential equations in finite element analysis and other simulation methods.

In summary, addition plays a vital role in engineering as a core tool in the design, analysis, and performance evaluation of systems and components across various disciplines. Its application ensures that systems are safe, efficient, and meet required performance standards.

Economics and Finance

In economics and finance, addition is essential for calculating profits, losses, interest rates, investments, and budgets. It is used to analyze economic and financial data and make informed decisions.

In the fields of economics and finance, addition plays a fundamental role in a wide range of applications. Here are more details on how addition is used in this context:

Profit and Loss Calculations: Businesses and individuals use addition to calculate profits and losses. This involves adding up revenues and subtracting costs and expenses to assess profitability.

Interest Rates and Financing: Addition is used to calculate interest, whether earned on investments or paid on loans. Summing up interest helps people and businesses understand how much they are gaining or paying.

Investments and Portfolio Performance: Investors use addition to calculate the total return on investments by summing the returns from different assets in a portfolio.

Personal and Business Budgeting: Addition is essential for creating and managing budgets. By adding income and expenses, individuals and companies can control their finances and make informed spending and investment decisions.

Financial Data Analysis: Addition is used in financial analysis to compute financial ratios and indicators such as liquidity ratios, debt ratios, and other measures that help assess a company's financial health.

Present and Future Value: In asset valuation and investment project assessments, addition is used to calculate net present value (NPV) and future value. These calculations are crucial for making investment decisions.

Accounting: In accounting, addition is used to keep accurate records of financial transactions by summing incomes, expenses, and adjustments.

National Economy: In economics, addition is used to calculate macroeconomic indicators such as gross domestic product (GDP), which involves adding up the value of all goods and services produced.

3.Subtraction: The operation of taking one number away from another

Subtraction is a mathematical operation used to find the difference between two numbers. It is represented by the symbol "−" and is performed by subtracting one number (called the subtrahend) from another larger number (called the minuend) to obtain the result, which is called the difference or remainder. For example, if we subtract 5 from 8, the calculation looks like this:

$8 - 5 = 3$

In this case, 8 is the minuend, 5 is the subtrahend, and the result, 3, is the difference. Subtraction is commonly used in a variety of situations to calculate changes, find missing amounts, or solve both mathematical and everyday problems.

Subtraction is a widely used mathematical operation in many everyday and academic contexts.

Mathematical Problems

Subtraction is essential in solving mathematical problems involving quantity comparisons. For instance, subtraction problems can be used to calculate age differences, distances, or time intervals.

Subtraction is crucial for solving mathematical problems that involve comparing quantities or determining differences. It is often used to answer questions like "What is the difference between...?" or "How much more/less...?"

Age Problems

In problems involving ages, subtraction is used to calculate the age difference between two people. For example: "If John is 35 and Mary is 28, what is the age difference between them?"

Distance Problems

To solve distance-related problems, subtraction is used to calculate the length or distance between two points in a straight line or along a road. For example: "If one city is 120 kilometers north of another, what is the distance between them?"

Time Problems

Subtraction is used in time problems to calculate the duration of events or the time difference between two moments. For example: "If a train leaves at 9:15 a.m. and arrives at 2:45 p.m., how long was the trip?"

Money Problems

In money-related problems, subtraction is used to calculate change from a purchase or to find the difference between income and expenses. For example: "If you buy an item for $25 and pay with $50, how much change do you receive?"

Quantity or Volume Problems

In problems involving quantities or volumes, subtraction is used to determine the difference in amounts. For example: "If you have 15 liters of water in a container and pour out 7 liters, how much is left in the container?"

Temperature Problems

Subtraction is applied to problems involving temperature to calculate the difference between two temperatures. For example: "If the current temperature is 28°C and the day's maximum was 35°C, what is the temperature difference?"

These are just a few examples of how subtraction is used in mathematical problems to compare quantities and calculate differences across a wide variety of situations. Subtraction is an essential tool in problem-solving and is applicable in many areas of mathematics and everyday life.

Accounting

In accounting, subtraction is used to calculate profits, losses, and balances. For instance, subtracting expenses from income determines net profit.

Subtraction plays a key role in determining profits and losses and evaluating financial balances.

Income and Expenses Calculation

Accounting involves recording income (like sales and other earnings) and expenses (such as operational costs, taxes, and financial expenses) of an entity, whether a business or an individual.

Net Profit Determination

To calculate net profit or earnings, the total expenses are subtracted from the total income. This shows how much an entity has earned after covering all expenses. The basic formula is:

Net Profit = Total Income − Total Expenses

Loss Identification

If expenses exceed income, a loss occurs. The loss is calculated by subtracting total income from total expenses, resulting in a negative number.

Balance Evaluation

In accounting, accounts are used to track financial transactions. Accounts may have debit balances (positive) or credit balances (negative). Subtraction is used to determine the account balance.

Budgeting and Financial Planning

Subtraction is applied in budget preparation and financial planning. By comparing budgeted income with budgeted expenses, one can determine whether a plan is profitable or if there is a budget deficit.

Audit and Account Review

In financial auditing, auditors review an entity's accounting records to ensure that income and expenses are properly recorded and that additions and subtractions are performed correctly.

Financial Statement Preparation

Accounting is used to prepare financial statements such as the balance sheet, income statement, and cash flow statement. These reports often involve addition and subtraction calculations to provide a comprehensive view of an entity's financial status.

Accounting relies on principles and techniques that involve subtraction to evaluate profitability, maintain accurate records, and ensure the integrity of financial reports. The ability to perform subtraction is essential to properly understand and manage the financial status of any entity, whether a company, nonprofit, or individual.

Inventory

Subtraction is applied to track inventory. By subtracting the quantity sold from the starting quantity, one can determine the stock available.

Subtraction plays an essential role in inventory management, enabling accurate tracking of stock and determining how many products or items remain available for sale.

Stock Control

Businesses and retailers use subtraction to track the quantity of goods available in their inventory at any given time.

Inventory Updates

After each sale, the number of products sold is subtracted from the starting quantity or current inventory balance. This is done to keep precise records of what remains in stock.

Reordering Inventory

Subtraction is also used to determine when inventory needs to be restocked. When the inventory balance reaches a predefined minimum level (reorder point), a new order is placed to avoid future shortages.

Loss and Theft Calculation

Subtraction is used to calculate inventory losses or theft. By comparing the initial quantity with the actual stock, any discrepancies can be identified and addressed.

Perishable Product Control

In managing perishable products, such as fresh food, subtraction is used to calculate how many items have expired or deteriorated. This helps prevent selling spoiled goods.

SKU Management (Stock Keeping Units)

Subtraction is applied in managing SKUs to ensure individual product units are tracked accurately. By subtracting sold or moved units, an accurate record of the location and availability of each SKU is maintained.

Inventory Optimization

Subtraction is used to evaluate inventory turnover and identify slow-moving products. This helps optimize inventory and minimize storage costs.

Accurate inventory management is essential to ensure businesses can meet customer demand and avoid financial losses from obsolete or out-of-stock products. Subtraction plays a critical role in this management by allowing companies to constantly monitor their stock and make informed decisions about restocking and stock control.

Changes and Variations

Subtraction is used to calculate changes and variations in data. For example, one can subtract the current temperature from a previous temperature to determine the temperature variation.

Subtraction is a useful tool for calculating changes and variations in a wide range of data. By subtracting a previous value from a current one, one can determine how much a quantity has changed over time or between two points.

Climatology

In climatology, subtraction is used to calculate changes in temperature, precipitation, or other climate variables. Subtracting the current temperature from the previous one provides information about temperature variation over a given period.

Price Variation

In economics and finance, subtraction is used to calculate the price variation of goods or financial assets. Subtracting the current price from the previous price gives the change in value.

Value Variation

In data analysis, subtraction is used to calculate value variation in a data series. For example, subtracting one month's value from the previous month's yields the monthly variation.

Inventory Variation

In inventory management, subtraction is used to calculate stock variation. Subtracting the previous inventory balance from the current balance shows how many items have been sold or added.

Difference Calculation

In various fields, subtraction is used to calculate measurement differences, such as elevation differences in topography or stock level differences in logistics.

Scientific Data Variation

In scientific research, subtraction is used to calculate changes in experimental data, such as the difference in chemical concentration before and after an experiment.

Trend Analysis

Subtraction is used to assess trends over time. Subtracting values at different times helps identify patterns and changes in data.

Coordinate Differences

In mathematics and geometry, subtraction is used to calculate differences in spatial or geographic coordinates, such as the difference in latitude and longitude between two points on Earth.

Subtraction is a fundamental operation for calculating changes and variations in data, whether in scientific, economic, or climate contexts, or any other field that requires measuring and understanding differences between values at different times or locations. This tool is essential for data analysis and identifying patterns and trends.

Personal and Business Budgeting

Subtraction is essential for creating and managing budgets. By subtracting expenses from income, one can assess whether a budget is balanced or if there is a deficit.

Subtraction plays a crucial role in the creation and management of budgets, both personal and business. It helps assess financial health and determine whether a budget is balanced, resulting in a surplus or indicating a deficit.

Personal Budgeting

Income and Expense Tracking: In a personal budget, subtraction is used to calculate the difference between income (such as salary, investment income) and expenses (such as housing, food, transportation, entertainment). Subtracting expenses from income determines whether there is a surplus or a budget deficit.

Budget Balance: If expenses are lower than income, the result is a budget surplus, indicating one is living within their means. If expenses exceed income, a deficit occurs, indicating the need to adjust the budget.

Financial Planning: Subtraction is used to assess the feasibility of financial goals, such as saving for an emergency fund, paying off debt, or investing for the future.

4.Multiplication: Repeating an addition several times, like 3 x 4 is the same as adding 3 four times (3 + 3 + 3 + 3)

Multiplication is a fundamental mathematical operation that consists of repeating a sum or addition several times. In other words, it is a shorthand way of adding a number a certain number of times. The symbol used to represent multiplication is "x".

For example, if we have the expression 3 x 4, this means that we are multiplying 3 by 4. To perform this multiplication, we add 3 four times:

3 x 4 = 3 + 3 + 3 + 3 = 12

So, 3 x 4 is equal to 12. Multiplication is an essential arithmetic operation and is used in a variety of contexts to calculate quantities, areas, volumes, and more. It can also be seen as an efficient way of counting repeated elements.

Multiplication is an essential mathematical operation that has wide applicability in daily life. It can be better understood by considering practical examples of how it is used in everyday situations:

Shopping at the supermarket: When you buy groceries, products often have a price per unit (for example, the cost of a can of soda). If you want to buy multiple cans, multiplication allows you to quickly calculate the total cost. If each can of soda costs $1 and you want to buy 6 cans, you can multiply the unit price by the quantity: 1 x 6 = $6.

Calculating time and speed: If you are traveling by car and want to know how long it will take to reach your destination at a certain constant speed, you can use multiplication. For example, if you are traveling at a speed of 60 kilometers per hour and want to know how long it will take to cover 120 kilometers, you can multiply the speed by the time: 60 km/h x 2 h = 120 km.

Area of a room: Imagine you want to buy carpets for a rectangular room. To determine how many square yards of carpet you need, you multiply the length by the width of the room. If the room is 10 feet long and 8 feet wide, the area is 10 feet x 8 feet = 80 square feet.

Cost calculations for events: If you are organizing an event and need to calculate the total cost of tickets, you can multiply the unit price by the number of tickets you plan to sell. For example, if a ticket costs $20 and you plan to sell 100 tickets, the total cost would be 20 x 100 = $2,000.

Cooking and recipes: When cooking, you often need to adjust the quantities of ingredients in a recipe. Multiplication allows you to scale the quantities of ingredients based on the number of servings you want to prepare. If a recipe requires 2 cups of flour and you want to double the recipe, you will need 4 cups of flour.

Salaries and hourly pay: In work situations, multiplication is used to calculate total salary. For example, if you earn $15 per hour and work 40 hours a week, your weekly salary would be 15 x 40 = $600.

Scheduling and organization: If you need to plan your day and allocate time to different tasks, you can use multiplication to estimate how much time you need for each activity. For example, if you plan to study for 2 hours and each study session is 30 minutes, you will need to study 2 x 2 = 4 sessions of 30 minutes.

Resource distribution: Multiplication is also used in the distribution of resources in contexts such as agriculture or construction. If a farmer plants corn on 5 hectares of land, they must calculate how many corn seeds are needed per hectare and then multiply the required amount by 5.

Multiplication is an essential mathematical tool that simplifies calculations in numerous aspects of daily life, from shopping and time planning to financial calculations and organization. It facilitates problem-solving and efficient decision-making in a wide variety of situations.

Multiplication is one of the fundamental mathematical operations that allows us to simplify calculations and perform everyday tasks more efficiently. Whether we are managing our finances, planning activities, cooking, organizing events, or solving problems, multiplication is an essential tool that helps us save time and effort.

By understanding how to apply multiplication in practical situations, we can make more informed decisions and perform calculations more accurately. This mathematical skill is not only useful in daily life, but it is also essential in fields such as science, engineering, economics, and many other disciplines where quantitative analysis is required.

Multiplication is a tool that allows us to approach a wide variety of tasks and problems more efficiently, enhancing our ability to make decisions and achieve our goals in everyday life. Whether we are solving math problems, planning activities, managing our finances, or performing everyday tasks, multiplication simplifies calculations and allows us to make informed decisions. Its usefulness extends to various fields of life, making it a fundamental skill in education and real-world problem solving.

Education: Multiplication is a basic skill taught in school and lays the foundation for more advanced mathematical concepts. It is a fundamental part of the school curriculum and is used to develop essential math skills in students of all ages.

Development of Mathematical Skills: Multiplication is a key starting point for the development of more advanced mathematical skills. Mathematical concepts such as division, geometry, and algebra often build on a solid understanding of multiplication.

Problem Solving: Multiplication is used to solve a wide variety of mathematical and real-world problems. Students learn to apply multiplication concepts to address practical situations and make informed decisions.

Applicability in Various Fields: The concepts of multiplication extend to fields beyond mathematics. Students acquire a skill that is relevant in science, technology, engineering, economics, and many other disciplines.

Improvement of Logic and Organization: Multiplication fosters logical thinking and organization. Students must follow specific steps and rules to perform multiplication calculations, which strengthens their reasoning ability.

Preparation for Daily Life: Knowledge of multiplication is practical in everyday life. Students learn to handle financial transactions, make purchases, plan tasks, and manage resources.

Facilitation of Abstract Concepts: Multiplication is often taught with concrete representations, such as groups of objects or arrangements. This helps students understand abstract concepts and visualize mathematical operations.

Stimulation of Logical Problem Solving: Multiplication problems challenge students to think logically and seek solutions. This fosters problem-solving and critical thinking.

Multiplication is a fundamental pillar in mathematical education. It helps students develop essential math skills, prepare for more advanced mathematical challenges, and acquire a tool that is relevant in their daily lives and various disciplines. Teaching multiplication is crucial for academic success and the development of problem-solving skills.

Science and Technology: In scientific fields like physics, chemistry, and biology, multiplication is applied in scientific calculations, such as determining concentrations, analyzing experimental data, and modeling natural phenomena.

Multiplication plays a fundamental role in science and technology. Here are some examples of how multiplication is applied in these fields:

Physics: In physics, multiplication is used in a variety of calculations. For example, to calculate speed, acceleration is multiplied by time ($v = at$). It is also used in Hooke's law, which relates the force applied to a spring with its elongation. Additionally, in mechanics, it is used to calculate the work done by a force on a moving object.

Chemistry: Multiplication is essential in chemical calculations, especially in stoichiometry. It is used to balance chemical equations, calculate the molar mass of compounds, determine the amount of reactants needed, and predict the amount of products that will form in a chemical reaction.

Biology: In biology, multiplication is applied in various areas. For example, it is used to calculate the growth rate of populations, determine the concentration of solutions in molecular biology labs, and quantify data in genetic and epidemiological studies.

Experimental Data: In the collection and analysis of experimental data, multiplication is used to calculate statistical measures, such as the mean, standard deviation, and variance. These measures are fundamental for summarizing and analyzing experimental results.

Modeling and Simulations: Multiplication is employed in modeling natural phenomena and in numerical simulations. For example, in climatology, equations involving multiplications are used to predict climate patterns and long-term changes.

In all these fields, multiplication is an essential tool that enables scientists to perform precise and quantitative calculations, which are fundamental for research, problem-solving, and decision-making in scientific and technological contexts. Mathematics, including multiplication, is the common language that drives progress in these fields.

Engineering: Engineers use multiplication to design and analyze structures, systems, and circuits, calculate loads and dimensions, and model the behavior of complex systems.

Multiplication plays a crucial role in engineering and is a fundamental tool in various engineering disciplines.

Structural Design: Civil and structural engineers use multiplication to calculate loads and stresses in structures, such as bridges and buildings. It is also applied in the design of connections and construction materials.

Mechanical Engineering: In mechanical engineering, multiplication is used in the design of machines and mechanical systems. For example, the forces and torques needed for a machine to operate correctly are calculated.

Electrical Engineering: In electrical engineering, multiplication is applied to calculate currents, voltages, and power in electrical circuits. It is also used in the design of circuits and control systems.

Chemical Engineering: Chemical engineers employ multiplication in calculations related to chemical processes, such as determining reaction rates, the amount of reactants needed, and the production of chemical products.

Software Engineering: In software engineering, multiplication is used in performance calculations, such as estimating execution times and planning resources for software development projects.

Aerospace Engineering: Aerospace engineers apply multiplication in the design calculations for aircraft and rockets, including determining the forces needed for takeoff and flight.

Modeling and Simulation: Multiplication is used in simulations and numerical models to predict the behavior of complex systems in engineering, such as fluid dynamics in aerodynamics.

Systems Engineering: Systems engineering uses multiplication in the analysis and design of complex systems where multiple components interact with each other.

Multiplication is an essential tool in engineering and is applied in a wide variety of disciplines to perform calculations, design systems and structures, model behaviors, and solve engineering-related problems. It is a fundamental mathematical skill for engineers in their daily work.

Economics and Finance: Multiplication is essential in calculating interest, profits, losses, exchange rates, and other financial aspects. It is used in investments, budgets, and financial analysis.

Interest: In calculating interest, multiplication is used to determine how much will be earned or paid in interest on loans, investments, or savings accounts. The basic simple interest formula involves multiplying the interest rate by the principal and the time in years.

Profits and Losses: In financial analysis, multiplication is applied to calculate profits or losses on investments or in the buying and selling of assets. The difference between the selling price and the buying price is multiplied by the quantity of assets.

Exchange Rates: In international trade and foreign exchange markets, multiplication is used to convert one currency into another using exchange rates. The amount of foreign currency is multiplied by the exchange rate to obtain the equivalent in local currency.

Budgets: In financial planning and budgeting, multiplication is applied to estimate expenses, revenues, and benefits based on various variables, such as the number of units sold and unit prices.

Investments: Investors use multiplication to calculate the return on their investments. This involves multiplying the amount invested by the expected rate of return to determine potential profits.

Loan Amortization: In the context of loans, multiplication is used to calculate periodic amortization payments. This involves multiplying the loan amount by the interest rate and a factor that accounts for the payment period.

Cost Analysis: Businesses use multiplication to analyze production costs and determine the total cost of manufacturing products. This includes calculating the cost of materials, labor, and other expenses.

Present and Future Value: Multiplication is fundamental in calculating the present and future value of cash flows. It allows for determining the current or future value of sums of money based on interest rates and time periods.

Multiplication plays an essential role in economics and finance by enabling financial calculations, estimating profits and losses, evaluating investments, and conducting financial analysis. It is a fundamental tool for individuals, businesses, and financial analysts making decisions related to money and economic transactions.

Times and Planning: In daily life, multiplication is applied in planning schedules and tasks. It helps estimate the time and resources needed to complete projects and activities.

Schedules and Programming: Multiplication is used to schedule activities and events in a timetable. For example, when planning a work meeting, you can multiply the estimated duration by the amount of available time to find a suitable time slot.

Duration of Activities: When planning projects or tasks, multiplication is employed to estimate how long it will take to complete each activity. Multiplying the number of time units (such as hours or days) by the number of activities provides an estimate of the total time required.

Resource Estimation: In project planning, multiplication is used to estimate the amount of resources needed, such as people, materials, or equipment. For example, if a certain number of structures need to be built, you can multiply the number by the amount of resources needed for each one.

Time Budgeting: Multiplication is applied in time budgeting. This involves calculating the total time necessary to carry out a project and ensuring that it fits within the scheduled deadlines.

Travel Planning: When planning a trip, multiplication is used to estimate the duration of each leg of the journey, such as flight time or driving time. This is crucial for scheduling activities and booking accommodations.

Personal Time Management: In everyday life, multiplication helps people manage their personal time. For example, when scheduling a daily routine, the hours assigned to each activity are multiplied by the number of days in a week.

Event Planning: Multiplication is applied in organizing events, such as weddings or parties. It helps estimate the amount of food, drink, or decorations needed based on the number of guests.

Cost Calculation: In project or event planning, multiplication is used to calculate time-related costs, such as worker salaries or facility rentals.

Multiplication is an essential tool for planning and time management in daily life. It allows people to estimate times, resources, and costs efficiently, which is fundamental for organizing activities, projects, and events. Accurate time planning is key to achieving goals and maximizing efficiency in everyday life.

Statistics and Data Analysis: Multiplication is used in statistics to calculate cross-products and in data analysis to perform matrix operations and model relationships between variables.

Multiplication plays a fundamental role in statistics and data analysis, as it is used in a variety of calculations and operations related to processing and interpreting data.

Cross Products in Contingency Tables: In statistics, contingency tables are used to show the relationship between two categorical variables. Multiplication is applied to calculate cross products, which represent the joint frequency of two specific categories in the variables.

Matrix Operations: In more advanced data analysis, matrices are used to represent and analyze multidimensional datasets. Matrix multiplication is employed in matrix operations, such as multiplying data matrices by coefficient matrices in regression analysis or in principal component calculations.

Modeling Relationships Between Variables: Multiplication is used in building statistical and mathematical models to describe relationships between variables. For example, in a linear regression model, coefficients are multiplied by the values of predictor variables to predict an outcome variable.

Calculating Descriptive Statistics: In descriptive statistics, such as calculating the weighted mean, multiplication is used to assign weights to values and calculate measures of central tendency.

Calculating Measures of Dispersion: Multiplication is applied to calculate measures of dispersion, such as variance and standard deviation. In these calculations, the differences between values and the mean are squared and then summed.

Calculating Joint Probabilities: In probability statistics, products are used in calculations of joint probabilities. The joint probability of two events is calculated by multiplying the probabilities of each individual event.

Applications in Multivariate Analysis: In multivariate analysis, matrix operations and calculations involving matrix multiplication are performed to explore relationships between multiple variables.

Simulation and Statistical Modeling: In statistical simulation and event modeling, Monte Carlo techniques that involve multiple repetitions of experiments are used. Multiplication is applied in these repetitive calculations.

Multiplication plays an essential role in statistics and data analysis by enabling calculations, modeling relationships, and processing information efficiently. It is a powerful tool in data-driven decision-making and understanding patterns and relationships in datasets.

Health and Medicine: In the field of medicine, multiplication is used in medication dosages, patient dose calculations, test result analyses, and medical research.

Multiplication is a mathematical operation that plays a crucial role in the field of medicine and health. It is used in various contexts to calculate medication dosages, perform test result analyses, engage in medical research, and manage healthcare.

Medication Dosage Calculations: In administering medications, multiplication is used to determine the exact amount of a medication that a patient should receive based on factors such as the patient's weight and the concentration of the medication. This ensures that the correct and safe dose is administered.

Medication Dilution: In some cases, medication dilutions are needed to adjust concentration. Multiplication is used to calculate the amount of diluent solution required to achieve the desired concentration.

Test Result Analysis: In clinical laboratories, test results are expressed in specific measurement units. Multiplication is applied to convert these results into understandable units or to perform calculations related to diagnoses and treatments.

Calculating Indices and Ratios: In medical research and epidemiology, multiplication is used to calculate indices and ratios that are fundamental for evaluating disease prevalence and treatment efficacy.

Estimating Survival Rates: In survival and prognosis studies, multiplication is employed to calculate survival rates and probabilities of specific events in patient populations.

Estimating Risks and Benefits: In clinical decision-making, multiplication is applied to assess the risks and benefits of different treatment options and medical procedures.

Pharmacokinetic Calculations: Multiplication is used in pharmacokinetics to model the distribution and elimination of medications in the body and to calculate key pharmacokinetic parameters, such as a drug's half-life.

Mathematical Modeling in Research: In medical and scientific research, mathematical models that include equations with multiplication are employed to understand and predict biological and medical phenomena.

Healthcare Management: In healthcare management and resource planning, multiplication calculations are used to estimate the demand for health services and allocate resources efficiently.

Multiplication is an essential tool in medicine and health, ensuring accuracy in medication administration, interpreting test results, and making clinical decisions. It is also an integral part of medical and scientific research, contributing to the advancement of understanding and treatment of diseases and disorders.

Business and Commerce: Businesses use multiplication to calculate profits, costs, profit margins, and sales estimates. It is also applied in financial analysis and business decision-making.

Multiplication plays a fundamental role in the world of business and commerce, as it is used in a wide variety of applications to perform financial calculations, estimate profits, costs, and make informed business decisions. Here are some examples of how multiplication is applied in this area:

Calculating Profits and Losses: In business operations, multiplication is used to calculate profits or losses by subtracting the cost of goods or services sold (COGS) from the selling price. This operation is essential for assessing a business's profitability.

Estimating Sales and Revenues: Multiplication is applied to estimate future sales and revenues based on the quantity of products or services sold and their selling price.

Calculating Taxes and Rates: In accounting and finance, multiplication is used to calculate sales taxes, interest rates, and other financial charges.

Budgets and Financial Projections: Businesses create budgets and financial projections that involve multiplication calculations. This includes estimating expenses, revenues, profit margins, and future cash flows.

Profit Margin Analysis: Multiplication is employed to calculate profit margins, which are the difference between the selling price and the costs associated with producing or acquiring goods and services.

Production and Manufacturing Costs: In manufacturing and production, multiplication calculations are used to determine production costs, including labor, materials, and other resources.

Interest and Financing Calculations: Multiplication is applied in calculating interest on loans, investment financing, and analyzing financing options.

Earnings Per Share Calculation: In financial analysis, multiplication is used to calculate earnings per share (EPS), an important indicator for stock investors.

Risk and Return Evaluation: Multiplication is employed in financial analysis to estimate potential returns and the risks associated with different investments and projects.

Cost-Benefit Analysis: In business decision-making, multiplication is applied in cost-benefit analysis to evaluate whether the expected benefits of an investment outweigh the associated costs.

Inventory Management: Multiplication is used in inventory management to calculate the amount of stock needed and to estimate the value of inventory at any given time.

Multiplication is an essential tool in the world of business and commerce, enabling companies to perform accurate financial calculations, assess profitability, and make data-driven decisions. It facilitates financial planning, resource management, and project evaluation, contributing to the success and growth of organizations.

Event Planning: When organizing events, multiplication is employed in budget calculations, attendee estimates, and the management of necessary resources.

Event Budgeting: Multiplication is used to calculate the estimated costs of an event. This includes estimating expenses for items such as venue rental, catering, decorations, entertainment, security, and other services. By multiplying the unit cost by the necessary quantity of each item, the total event cost is obtained.

Estimating Attendees: Multiplication is employed to estimate the number of attendees expected at an event. This is essential for planning logistics, such as the number of chairs, food, and drinks needed, and the layout of the venue.

Calculating Time and Resources: When planning an event, it's important to determine how much time and resources will be needed to carry out various activities. Multiplication is used to estimate how long specific tasks will take and to allocate resources, such as personnel and equipment.

Scheduling Activities: Multiplication is applied in scheduling activities to ensure that the event's activities are executed at the right times. For example, determining how much time is needed for each presentation or activity and then scheduling them in sequence.

Calculating Space and Capacity: If the event is to be held in a venue with limited capacity, multiplication is used to determine how many people can be safely accommodated. This is essential for ensuring the safety and comfort of attendees.

Distribution of Materials: In events involving the distribution of promotional materials, such as brochures or samples, multiplication is applied to calculate the amount of materials needed and to ensure that there are enough for all attendees.

Food and Beverage Planning: If the event includes food and beverages, multiplication is used to calculate the amount of food and drink needed, as well as to plan the number of tables and chairs required.

Cost Per Attendee Evaluation: Multiplication is applied to calculate the cost per attendee, which is useful for determining ticket prices or assessing the event's profitability.

Revenue and Expense Management: In the financial management of the event, multiplication is used to calculate expected revenues based on ticket sales, sponsorships, and other factors, as well as to estimate total expenses.

Multiplication is a fundamental tool in event planning, allowing organizers to perform accurate calculations to ensure that the event runs efficiently and successfully. It facilitates cost estimation, resource management, and informed decision-making throughout all stages of event planning and execution.

Multiplication is an essential mathematical tool that not only is an integral part of education but also is a fundamental skill in practically all aspects of life and in numerous professional fields. It facilitates calculations, analysis, and decision-making, making it a crucial component in real-world problem-solving.Division is a mathematical operation used to distribute or share a number into equal parts. In other words, division is performed to determine how many times one number (the dividend) fits into another larger number (the divisor) in an equitable manner. The result of a division is called the quotient.

5.Division: Splitting a number into equal parts, like dividing 12 into 4 equal groups results in 3 in each group (12 ÷ 4 = 3)

Division is a mathematical operation used to divide or share a number into equal parts. In other words, division is performed to determine how many times one number (the dividend) fits into another, larger number (the divisor) in an equal manner. The result of a division is called the quotient.

The example you mentioned, "12 ÷ 4 = 3," illustrates how division works. Here is an expanded explanation:

Dividend: The number to be divided is called the dividend. In this case, the dividend is 12.

Divisor: The number by which the dividend is divided is called the divisor. In this case, the divisor is 4.

Quotient: The result of the division is the quotient. In the example, the quotient is 3.

So, when you divide 12 by 4, you are distributing 12 into 4 equal groups. This means that each group will receive 3 as part of the distribution. It is important to remember that division seeks to ensure that the amount is evenly divided among groups or parts.

Division is used in a variety of situations in everyday life and in mathematics. For example, it is used to calculate the number of people in each row of seats on a bus, to divide a pizza into equal slices, to distribute a budget across various expenses, to calculate rates, and to provide solutions to mathematical and practical problems that require the equitable distribution of resources or quantities.

Division is an essential mathematical operation that is applied in various situations both in everyday life and in mathematics and other disciplines. Here are additional examples of how division is used in different contexts:

Resource Allocation in Business: In the business realm, division is used to distribute resources, such as the marketing budget, equitably among different areas or projects. The allocation of resources in the business context is a common practice to ensure an efficient and fair distribution of available resources, such as the marketing budget. Division plays a crucial role in this process by allowing companies to allocate resources fairly among different areas, projects, or initiatives.

Budget Allocation: The budget is a critical resource for any business. Division is used to divide the marketing budget into equal parts or specific proportions for different areas or projects. This ensures that each area or project receives a fair share of the available budget.

Resource Optimization: Division is not limited to equal distribution; it can also be used to allocate resources in a way that maximizes return or investment. This involves assigning more resources to areas or projects with greater growth or profitability potential.

Strategic Planning: Resource division in the strategic planning process is essential to ensure that the company's goals and objectives are effectively met. Each project or area receives the necessary amount of resources to achieve its specific objectives.

Performance Evaluation: Division allows companies to compare the performance of different areas or projects. By allocating resources equitably or based on merit, outcomes can be evaluated, and informed decisions can be made regarding the redistribution of resources in the future.

Staff Distribution: Beyond budgeting, division applies to personnel distribution. Teams and employees are assigned to specific projects or areas according to their skills and available human resources.

Cost-Benefit Evaluation: Resource division is also used in cost-benefit evaluation. It allows for determining how much is being invested in a project or area in relation to the results obtained.

Data-Driven Decision Making: Data and analysis-based division help companies make informed decisions about resource allocation. This ensures that resources are used efficiently and that maximum value is obtained.

Division in resource distribution in business is a key tool for effectively managing available assets, whether financial, human, or otherwise. It facilitates strategic decision-making, cost control, and result optimization, which are essential for the success and sustainability of businesses.

Time Division: In time management, division is applied to schedule tasks or activities throughout the day. For example, if you have 8 working hours and want to divide them equally between two projects, you would work 4 hours on each.

Time division is a common practice in time management and activity planning. It is used to schedule tasks and activities throughout the day equitably, especially when multiple projects or responsibilities need attention. The goal of time division is to ensure that each task receives the necessary attention and that time is used efficiently.

Daily Task Planning: In time management, division is used to plan how working hours will be distributed throughout the day. This involves allocating time to specific tasks, projects, or activities.

Balance Between Projects: When you have several projects or responsibilities, division is applied to ensure that all receive attention. For example, if you have 8 working hours and want to divide them between two projects, you would work 4 hours on each. This helps maintain a balance among different tasks.

Work and Rest Intervals: Time division also applies to scheduling work and rest intervals. Setting scheduled breaks can improve productivity and reduce fatigue.

Task Prioritization: When dividing time, it is important to prioritize tasks and activities. Some tasks may require more time than others, so it is essential to allocate time based on the importance and urgency of each task.

Meeting and Commitment Management: Time division is used to schedule meetings, appointments, and other commitments. This helps ensure that time is available and used efficiently to meet these commitments.

Progress Evaluation: Time division allows for tracking progress on tasks and projects. By comparing the time spent with the results achieved, areas for improvement or adjustments in planning can be identified.

Flexibility: Although division is a useful tool, it is also important to be flexible and adapt to changing circumstances. Sometimes, it is necessary to adjust the division of time based on emerging needs.

Time division is an effective strategy for organizing tasks and projects. It helps maintain balance in time management, which is essential for productivity and efficiency at work. Additionally, it allows individuals to effectively meet multiple responsibilities and commitments, ensuring that time is utilized optimally.

Food Distribution: Division is employed in the food industry to determine food portions in containers or plates. For example, dividing a can of soup into 4 equal bowls.

Division is fundamentally applied in the food industry to determine food portions in containers or plates. This practice is essential to ensure that food is distributed evenly and that portions meet the required standards of quality and quantity.

Food Packaging: In food production and packaging, division is used to determine the amount of food placed in each container. For example, dividing a can of soup into a specific number of individual containers or into smaller portions suitable for consumption.

Restaurants and Catering: Restaurants and catering services use division to ensure that dishes are served with uniform portions. This is essential to provide a consistent dining experience to customers and to control costs.

Calorie and Nutrient Control: Division is applied in nutritional labeling of processed foods. Food is divided into portions, and calories and nutrients per portion are calculated, allowing consumers to know the nutritional content of products.

Schools and Cafeterias: In school cafeterias and in the collective catering industry, division is used to serve meals equitably to a large number of people. The total amount of food is divided into individual portions to meet the needs of diners.

Plate Preparation: In the kitchen, division is applied to prepare individual plates with specific portions of each component. This is important to ensure that ingredients are evenly distributed on the plate.

Frozen Food Packaging: In the frozen food industry, food is divided into individual portions that are packaged and sold as ready-to-heat and serve meals.

Menu Portions: Restaurant menus often indicate food portions to help customers make informed decisions about what to order. Division is applied to determine the appropriate quantities of each dish.

Quality Control: Division is also an essential tool for quality control in the food industry. It ensures that products meet specified quality and quantity standards.

Division in food distribution ensures that food products are presented and served uniformly and consistently. This is important both to meet consumer expectations and to maintain quality and food safety in food production and service.

Proportions in Cooking and Baking: In cooking and baking recipes, division is used to adjust ingredient quantities and calculate the appropriate proportions.

Applying division in cooking and baking is fundamental to ensuring that recipes are accurate and that ingredients are used in the correct proportions. Precision in the division of ingredients is essential for the success of a recipe, as it affects the flavor, texture, and final outcome of dishes and desserts.

Recipe Size Adjustment: Division is applied to adjust recipe sizes. For example, if an original recipe is designed to serve 4 people, but you want to prepare it for 8 people, you need to divide the ingredient amounts by half to double the recipe.

Portion Calculation: Division is used to calculate the number of portions that can be obtained from a recipe. By dividing the total amount of food by the size of a portion, you get the number of servings the recipe will yield.

Ingredient Proportions: In each recipe, ingredients must be divided into specific proportions. For example, in baking, it is crucial to divide the amount of flour, sugar, butter, and other ingredients to ensure that the dough or mixture is balanced and produces a quality final product.

Measurement Conversions: Division is applied in converting ingredient measurements. For example, if a recipe requires 1 cup of flour and you want to know how many grams that is, you need to divide the number of cups by the gram equivalence.

Doubling or Reducing Recipes: When you want to double or reduce a recipe, division is used to adjust ingredient quantities proportionally.

Calculating Partial Ingredient Recipes: In many recipes, a part of an ingredient is used that is sold in a larger package (like a can of tomatoes or a carton of milk). Division is applied to calculate the necessary amount from the larger package.

Dough Portioning: In preparing dough, such as cookie or bread dough, division is used to divide the dough into equal portions before baking.

Correcting Measurement Errors: If a measurement error is made for an ingredient, division can be applied to correct the amount by adding or subtracting ingredients as needed.

Division in cooking and baking is an essential skill for chefs and bakers, ensuring that recipes are accurate and that food is prepared consistently. This is especially important in baking, where ingredient proportions have a critical impact on the quality and flavor of baked goods.

Unit Conversion: Division is applied in converting units of measurement, such as converting miles to kilometers, pounds to kilograms, or degrees Celsius to Fahrenheit.

Unit conversion is a common application of division and is essential in situations where it is necessary to switch from one unit of measurement to another. Division is used to establish a relationship or conversion factor between the two units and then applied to perform the conversion.

Miles to Kilometers: To convert miles to kilometers, a conversion factor of approximately 1.60934 is used (or 1 mile ≈ 1.60934 kilometers). To perform the conversion, you divide the number of miles by this conversion factor. Example: To convert 10 miles to kilometers, you perform the division: 10 miles ÷ 1.60934 ≈ 6.214 kilometers.

Pounds to Kilograms: To convert pounds to kilograms, a conversion factor of approximately 0.453592 is used (or 1 pound ≈ 0.453592 kilograms). To perform the conversion, you divide the number of pounds by this conversion factor. Example: To convert 20 pounds to kilograms, you perform the division: 20 pounds ÷ 0.453592 ≈ 44.092 kilograms.

Degrees Celsius to Fahrenheit: The conversion between degrees Celsius (°C) and degrees Fahrenheit (°F) involves a formula that includes division. The formula for converting from °C to °F is: °F = (°C × 9/5) + 32. In this formula, multiply the temperature in degrees Celsius by 9/5 and then add 32. Example: To convert 25°C to °F, you perform the division: (25°C × 9/5) + 32 ≈ 77°F.

Meters to Feet: To convert meters to feet, a conversion factor of approximately 3.28084 is used (or 1 meter ≈ 3.28084 feet). To perform the conversion, you divide the number of meters by this conversion factor. Example: To convert 5 meters to feet, you perform the division: 5 meters ÷ 3.28084 ≈ 16.404 feet.

Gallons to Liters: To convert gallons to liters, a conversion factor of approximately 3.78541 is used (or 1 gallon ≈ 3.78541 liters). To perform the conversion, you divide the number of gallons by this conversion factor. Example: To convert 10 gallons to liters, you perform the division: 10 gallons ÷ 3.78541 ≈ 37.854 liters.

Division is a key operation in unit conversion, as it allows for establishing proportional relationships between different unit systems. This is useful in a wide variety of contexts, from science and engineering to cooking and travel, where it is necessary to convert measurements to understand and effectively use different unit systems.

Statistics and Proportions: In statistics, division is used to calculate proportions, percentages, and rates. For example, to determine the unemployment rate, you divide the number of unemployed people by the total working-age population.

Division plays a fundamental role in statistics in calculating proportions, percentages, rates, and other important indicators that allow for data analysis and evidence-based decision-making. These calculations provide valuable insights into trends, relationships, and comparisons among different data sets.

Unemployment Rate: The unemployment rate is a key economic indicator that measures the percentage of unemployed individuals in relation to the labor force, i.e., those who are of working age and seeking employment. To calculate the unemployment rate, division is used as follows:

Divide the number of unemployed people by the total number of people in the labor force.

The result is multiplied by 100 to express the rate as a percentage. Formula: Unemployment Rate = (Unemployed Persons / Labor Force) x 100

For example, if there are 5,000 unemployed people in a city and the labor force in that city is 50,000 people of working age, the unemployment rate is calculated as follows: Unemployment Rate = (5,000 / 50,000) x 100 = 10% This rate of 10% indicates that 10% of the labor force in the city is unemployed.

Division is similarly applied in other areas of statistics, such as calculating gender proportions, growth rates, success percentages in experiments, mortality rates, interest rates, and more. These calculations allow analysts and statisticians to interpret data meaningfully and communicate results in a way that is easy to understand and compare. Division is an essential tool for statistical analysis and informed decision-making in various fields, from economics to public health.

Division of Resources in Households: In the domestic realm, division is applied to split responsibilities and tasks among family members or to distribute time and available resources equitably.

The division of resources and responsibilities in households is a common practice used to ensure a fair distribution of household duties, time, and available resources among family members. This is essential for maintaining balance in family life, ensuring that tasks are completed efficiently, and promoting cooperation among family members.

Household Tasks: At home, division is used to assign domestic tasks, such as cleaning, cooking, laundry, and gardening, among family members. This ensures that household responsibilities are shared fairly and that everyone contributes to the functioning of the home.

Child Care: In families with children, division applies to child care. Parents may divide their time and responsibilities for care, such as meal preparation, homework assistance, and transportation to extracurricular activities, so that both parents participate equitably.

Schedules and Activities: Division is used to plan family schedules and activities. This may include scheduling children's extracurricular activities, family quality time, and coordinating schedules to ensure everyone can participate in desired activities.

Use of Household Resources: Division of resources at home can apply to the use of appliances, kitchen utensils, storage space, and other shared resources. This prevents conflicts and ensures that resources are used equitably.

Family Finances and Budgeting: Division applies in managing household finances, where it is determined how family money will be spent, financial responsibilities are assigned, and budgets for various expenses are set.

Rest and Recreation Times: Division of rest and recreation time is important to ensure that all family members have the opportunity to relax and enjoy their free time. This may include taking turns watching television shows, family game time, and recreational activities.

Division in the domestic realm not only facilitates the equitable distribution of responsibilities but also promotes collaboration, communication, and mutual support among family members. This is essential for maintaining a harmonious and efficient environment at home. Additionally, division in the household can be adapted to the needs and preferences of each family and can evolve over time as circumstances change.

Division is a fundamental mathematical skill that allows us to divide quantities and resources fairly and efficiently. Whether solving everyday problems or performing more complex calculations in mathematics and other disciplines, division plays an essential role in the management and distribution of resources in our daily lives.

6.Fractions: Numbers like 1/2 or 3/4 that represent parts of a whole

Fractions are a way to represent numbers that describe parts of a whole. They consist of two parts: the numerator and the denominator. The numerator represents the number of parts being taken, and the denominator indicates how many parts the whole is divided into. To better understand fractions, here's a more detailed explanation along with examples:

Numerator and Denominator: In a fraction, the numerator is on the top and the denominator is on the bottom. For example, in the fraction 3/4, "3" is the numerator and "4" is the denominator. The numerator and denominator are two key components of a fraction that determine its value and meaning.

Numerator: The numerator is the number found at the top of a fraction. It represents the number of parts we are considering or taking from the whole. In the fraction 3/4, the numerator is "3," which means we are considering three parts of the whole.

Denominator: The denominator is the number located at the bottom of a fraction. It indicates how many equal parts the whole is divided into. In the fraction 3/4, the denominator is "4," meaning the whole has been divided into four equal parts.

So, in the example of the fraction 3/4:

The numerator (3) represents that we are considering three parts of something.

The denominator (4) indicates that the whole has been divided into four equal parts.

Together, the fraction 3/4 tells us that we are taking three of the four equal parts into which the whole has been divided. Fractions are a powerful way to express parts of a set, and the numerator and denominator are fundamental to understanding their meaning and value.

Parts of a Whole: Fractions are used to represent parts of an object, quantity, or set. For example, if we have a whole pizza and divide it into 4 equal parts, each of those parts can be represented by the fraction 1/4.

Imagine you have a whole pizza and you divide it into 4 equal parts, as if you were cutting the pizza into 4 identical pieces. Each of those pieces is considered an equal part of the whole, and you can represent each of them with the fraction 1/4.

The numerator, which is "1," represents how many of those equal parts you are considering. In this case, you are considering one of the four parts. The denominator, which is "4," indicates into how many equal parts the whole, which is the complete pizza, has been divided.

So, in simple terms, 1/4 of the pizza means you have one of the four parts into which the whole pizza was divided. This is a clear representation of how fractions are used to express parts of an object, quantity, or set, and how the numerator and denominator work together to describe this relationship. Fractions are a fundamental tool in mathematics and daily life for expressing proportions and divisions of a whole into smaller parts.

Proper and Improper Fractions: Proper fractions are those in which the numerator is less than the denominator. For example, 2/5 is a proper fraction because it represents less than half of the whole. Improper fractions have a numerator greater than or equal to the denominator, such as 7/4, which represents more than one whole unit.

Proper Fractions: Proper fractions are those in which the numerator is less than the denominator. This means they represent a quantity that is less than a complete unit or the whole they refer to. In other words, in a proper fraction, you are taking a part of a whole, but that part is less than the whole.

Example: The fraction 2/5 is proper because the numerator (2) is less than the denominator (5). It represents two parts of a set divided into five equal parts, which is less than half of the whole.

Improper Fractions: Improper fractions are those in which the numerator is equal to or greater than the denominator. This means they represent a quantity that is equal to or greater than a complete unit or the whole they refer to. In an improper fraction, you are taking an amount that exceeds the whole.

Example: The fraction 7/4 is improper because the numerator (7) is greater than the denominator (4). It represents seven parts of a set divided into four equal parts, which is more than one complete unit.

Proper fractions indicate that you are taking a part that is less than the whole, while improper fractions indicate that you are taking an amount that is equal to or greater than the whole. Both types of fractions are used in various situations, and it is important to understand their meaning and how they relate to the whole they reference.

Mixed Fractions: Mixed fractions combine an improper fraction with a whole number. For example, 1 1/2 is read as "one and a half" and represents 1 whole plus 1/2.

Mixed fractions are a way to express quantities that combine a whole number and an improper fraction. This notation is used to represent numbers more accurately and understandably in situations where values are not whole numbers but also not purely fractional.

Mixed Fraction: A mixed fraction consists of two parts: a whole number and an improper fraction. A space or a diagonal line is used to separate the whole number from the improper fraction. For example, 1 1/2 is read as "one and a half."

The whole number (in this case, "1") represents a complete quantity or whole units. The improper fraction (in this case, "1/2") represents a fractional amount or a part of the whole.

Example: If you have 1 1/2 apples, it means you have one whole apple (1) and half an apple more (1/2). In total, you have 1.5 apples.

Mixed fractions are useful in situations where it's necessary to express quantities that are not complete whole numbers but also do not fit well into a purely fractional representation. This notation makes it easier to understand partial quantities in everyday and mathematical contexts.

Examples of Fractions in Everyday Contexts:

Cooking Recipes: Cooking recipes often use fractions to indicate the amount of ingredients. For example, 1/2 cup of sugar means you should use half a cup.

Fractions are very common in cooking recipes and are an essential tool to accurately indicate the quantity of ingredients to be used when preparing a recipe. This is especially important in cooking, where proper proportions are crucial to achieving the desired flavor and texture in dishes.

Measuring Dry Ingredients: Fractions are used to measure dry ingredients like flour, sugar, salt, or spices. For example, "1/2 cup of flour" means you need to measure half a cup of flour.

Liquids: Fractions are also used to measure liquids like milk, oil, water, juice, etc. For example, "1/4 cup of oil" means you need to measure a quarter of a cup of oil.

Eggs: Often, recipes indicate fractions for the number of eggs to use. For example, "2/3 cup of beaten eggs" means you should beat eggs until they fill two-thirds of a cup.

Fruits and Vegetables: Fractions are also used to measure fruits and vegetables, especially when they are cut into pieces. For example, "1/4 cup of chopped tomatoes" means you need to chop enough tomatoes to fill a quarter cup.

Cooking Time: Although they are not fractions in themselves, cooking times are also a way to represent parts of a whole in recipes. For example, "bake at 350 degrees Fahrenheit for 30 minutes" indicates that you should bake for one-thirtieth of an hour.

The use of fractions in recipes is fundamental for cooks to follow instructions accurately and achieve the desired results in the kitchen. The ability to measure and use fractions effectively is an important skill for anyone who enjoys cooking or wants to successfully follow recipes.

Length Measurements: In a measuring tape, the marks represent fractions. If you are at 2 1/2 inches from the end of the tape, that means you are two and a half inches from the start.

In length measurements, fractions are commonly used in measuring instruments, such as measuring tapes, rulers, and scales, to indicate distances accurately. This makes it easier to measure lengths that are not complete whole numbers.

On a measuring tape, the marks are spaced in units of measurement, such as inches or centimeters. Each of those units can be subdivided into fractions, allowing for more precise measurements. For example, on a measuring tape in inches, the marks often include fractions such as 1/4, 1/2, and 3/4 of an inch.

Suppose you want to measure the length of an object and notice that the corresponding mark is between 2 and 3 inches on the measuring tape. If the mark is exactly halfway between 2 and 3 inches, that means you are at 2 1/2 inches from the end of the tape. In other words, you have measured a distance of two and a half inches from the beginning of the measuring tape.

Fractions in length measurements are essential for obtaining precise measurements, especially when you need to express distances that are not exact multiples of the unit of measure. This notation allows people to perform detailed measurements in everyday and professional situations, from construction and carpentry to sewing and crafting.

Hours in a Day: The day is divided into 24 hours. Each hour represents 1/24 of the day. If it is 3:30 PM, that means 3 and a half hours of the day have passed, or 3 1/2 hours.

The day is divided into 24 hours, which means each hour represents one twenty-fourth (1/24) of the entire day. This division into hours is standard in most parts of the world and is widely used in time measurement. Here's an example to illustrate how fractions of hours are used on the clock:

If the clock shows 3:30 PM, it means that 3 full hours have passed and also a fraction of an hour. To express this fraction of an hour in terms of fractions, it can be said that it is 3 hours and 1/2 hour, or 3 hours and 30 minutes. You can also express this in terms of fractions of an hour instead of minutes:

3 hours and 30 minutes can be expressed as 3 1/2 hours, meaning that three and a half hours of the day have passed.

This fraction notation is useful for representing exact times and understanding how many hours have elapsed since a reference point, such as midnight. Fractions of an hour are commonly used in everyday situations, such as scheduling meetings, calculating travel time, or keeping track of the time it takes to complete a task.

Dividing a Pizza: If you divide a pizza into 8 slices and take 3 of them, you can represent this as 3/8 of the pizza that you have eaten.

By dividing a pizza into 8 slices and taking 3 of those slices, you can represent the amount you have consumed as 3/8 of the pizza. This fraction notation is an effective way to express the proportion of the pizza you have eaten in relation to the whole. Here's a more detailed explanation:

Pizza Divided into 8 Slices: When the pizza is divided into 8 equal slices, each of those slices represents one-eighth of the whole pizza. Each eighth is equivalent to 1/8 of the pizza.

Taking 3 Slices: If you take 3 of those 8 slices, you are taking three-eighths of the pizza. Mathematically, this is represented as 3/8.

So when you say you have taken 3/8 of the pizza, it means you have consumed three of the eight parts into which the pizza was divided. This fraction notation allows you to accurately express the amount you have eaten concerning the whole. Fractions are useful in situations like this, where you want to express the part of a set or a whole clearly and concisely.

Purchases: If you have $20 and spend $4, you have spent 4/20 of your money, which can be simplified to 1/5, meaning you have spent one-fifth of your money.

When shopping and spending an amount of money, you can express the fraction of the money spent in relation to the total available. In your example, if you have $20 and spend $4, you can calculate the fraction of money spent in relation to the total available as follows:

Money Spent: You spent $4.

Total Available: You initially had $20.

To calculate the fraction of money spent in relation to the total available, you can express it as a fraction:

Money Spent / Total Available = $4 / $20

To simplify this fraction and express it more concisely, you can divide both the numerator and denominator by the greatest common divisor, which in this case is 4:

($4 ÷ 4) / ($20 ÷ 4) = $1 / $5

This reduces to the fraction 1/5. This means you have spent one-fifth of your available money. Fractions are useful for expressing proportions and relationships between amounts, which is especially helpful in financial situations and everyday spending.

Fractions are a versatile mathematical tool used in a variety of situations to describe parts of a set or a quantity. They are fundamental in everyday life and in fields such as mathematics, cooking, construction, finance, and more, where it is necessary to represent and work with parts of a whole.

7.Decimals: Numbers like 0.5 or 2.75 that represent a part of a whole based on 10

Decimals are a way of representing numbers that include fractions of a whole and are based on the decimal number system, which is related to base 10. Here is a more detailed explanation of decimals along with explanatory examples:

Decimal System and Base 10: The decimal system is the most widely used number system in the world and is based on the idea that each position in a number represents a power of 10. For example, in the number 123, the "1" represents one hundred (10^2), the "2" represents twenty (10^1), and the "3" represents three (10^0). The decimal system, also known as base 10, is the most commonly used numerical system worldwide. It is based on the idea that each position in a number represents a power of 10.

Base 10: The term "base 10" means that the system is based on the number 10. This means there are 10 different digits used to represent all numbers in this system: 0, 1, 2, 3, 4, 5, 6, 7, 8, and 9.

Decimal Positions: In the decimal system, each digit in a number has a specific position that represents a power of 10. Starting from the right, the first position after the decimal point is 10^0, the next position is 10^1, the next is 10^2, and so on. This is because each position is multiplied by 10 in relation to the previous position.

Example (123): To illustrate, the number 123 is broken down as follows:

The digit "3" is in the position 10^0, meaning it represents "three units."

The digit "2" is in the position 10^1, meaning it represents "two tens" (20).

The digit "1" is in the position 10^2, meaning it represents "one hundred" (100). So, when you add these parts, you get 100 (hundreds) + 20 (tens) + 3 (units), resulting in 123 in the decimal system.

The decimal system is fundamental in mathematics and everyday life, as it is used for counting, measuring, performing arithmetic operations, and expressing quantities in most situations. Understanding base 10 is essential for developing basic mathematical skills and for working with numbers in general.

Decimals as Fractions: Decimals are used to represent fractions of a whole unit. The decimal point (or comma) is used to separate the whole part of the number from the fractional part. For example, in the number 2.75, the "2" is the whole part, and the "0.75" is the fractional part that represents 75 hundredths of a unit.

Whole Part: The whole part of a decimal number is the part to the left of the decimal point. It represents the number of whole units present in the number.

Fractional Part: The fractional part of a decimal number is the part to the right of the decimal point. This part represents a fraction of a whole unit. The digits in the fractional part relate to negative powers of 10.

For example, in the number 2.75:

The "2" is the whole part, meaning you have two whole units.

The "0.75" in the fractional part represents 75 hundredths (or 75/100) of a whole unit. You can think of this as 3/4 of a unit. So, in this case, 2.75 represents 2 whole units plus 75 hundredths of a unit.

Decimals are a very useful way to express fractional quantities in various contexts, such as money (cents), measurements (e.g., 2.5 meters), or ratios (like 0.75, which is equal to 3/4). They allow for precise representation of values that are not complete whole numbers and are widely used in mathematics and everyday life.

Reading Decimals: When reading decimals, the digits after the decimal point are pronounced as if they were whole numbers. For example, 2.75 is read as "two point seventy-five."

Here are some additional examples of how decimals are read:

0.5: It is read as "zero point five," which is equivalent to "five tenths" or "half."

1.25: It is read as "one point twenty-five," which is equivalent to "one unit and twenty-five hundredths."

3.14159: It is read as "three point one four one five nine," which is an approximation of the number π (pi), with the digits after the decimal pronounced individually.

0.75: It is read as "zero point seventy-five," which is equivalent to "seventy-five hundredths" or "three quarters."

Reading decimals this way is common practice and helps express fractional quantities clearly and accurately in the decimal system. It facilitates communication in everyday situations and in mathematics, especially when dealing with quantities that are not complete whole numbers.

Explanatory Examples:

0.5: In this case, 0.5 represents half of a unit. If discussing money, this could be 50 cents.

2.75: This number represents two whole units plus 75 hundredths of a unit. You could think of this as 2 dollars and 75 cents.

3.14159: This number is an approximation of π (pi), an important mathematical constant. Here, 3 is the whole part, and the decimal digits represent fractions of π.

0.25: This decimal represents a quarter of a unit or 25 hundredths. If measuring time, this could be equivalent to a quarter of an hour, i.e., 15 minutes.

Decimals are a powerful tool for representing quantities that are not complete whole numbers and are widely used in mathematics, science, finance, and many other areas. They are particularly useful for expressing precise and fractional amounts of a unit or set.

Decimals are a fundamental tool in numerous fields due to their ability to represent precise and fractional quantities. Here are some areas where decimals play an important role:

Mathematics: Decimals are fundamental in arithmetic and algebra, as they allow for the exact expression of fractions and rational numbers. They are also essential in precision calculations, such as square roots, logarithms, and trigonometric equations.

Expression of Fractions and Rationals: Decimals are an effective way to express fractions and rational numbers in decimal form. This simplifies calculations and comparisons. For example, 1/2 is expressed as 0.5 in decimal notation, allowing for easier addition, subtraction, multiplication, or division of fractions.

Precision Calculations: Decimals allow for precision calculations, which are essential in mathematics and science. In situations where fractions may be difficult to manipulate, decimals simplify calculations. For example, when calculating the square root of a non-integer, decimals are used to express the answer accurately.

Logarithms and Exponential Functions: In algebra and trigonometry, decimals are crucial in calculations related to logarithmic and exponential functions. Logarithms and exponential functions are expressed in terms of decimal numbers to solve problems and model natural phenomena.

Rounding and Approximation: In mathematics, it is common to work with decimal numbers to round or approximate results. This is useful for simplifying answers and making calculations more manageable. For example, you might round 3.14159 to 3.14 to simplify calculations.

Scientific Notation: Decimals are used in scientific notation to represent very large or very small numbers concisely. This is common in mathematics and sciences, where extremely large values (like the distance between galaxies) or extremely small values (like the size of subatomic particles) are dealt with.

In summary, decimals are an essential tool in mathematics and are fundamental for performing a wide range of calculations and solving problems in algebra, calculus, trigonometry, and other mathematical areas. Their versatility and ability to express fractions and rational numbers accurately make them an integral part of mathematical work.

Science: In science, decimals are used to measure and express quantities precisely, such as distances, times, temperatures, concentrations, speeds, and other fractional values. They are also fundamental in scientific notation for representing very large or very small numbers.

Precise Measurements: In all scientific disciplines, precise measurements are made. Decimals allow for expressing measurements with a high degree of accuracy. For example, the length of a sample in chemistry, temperature in physics, or the concentration of a substance in biology can be expressed in decimal form to reflect exact measurements.

Units of Measure: Units of measure in science are often expressed in decimal notation. For example, the speed of light in a vacuum is expressed as approximately 299,792,458 meters per second. This implies a high degree of precision, and decimal notation facilitates the representation of these quantities.

Scientific Notation: In scientific notation, decimals are used to represent very large or very small numbers more compactly. This is essential in fields like astronomy (where astronomical distances are handled) and particle physics (where subatomic sizes are dealt with).

Precision Calculations: In scientific calculations, especially in chemistry, physics, and statistics, decimals allow for precision calculations and provide reliable results. This is important to ensure that scientific conclusions are robust.

Time and Durations: Decimals are used to express time and durations in seconds, minutes, hours, or even years with precision. For example, the average lifespan of a subatomic particle can be expressed in decimal form for accurate calculations.

Data Distribution: In science, decimals are used to represent the distribution of experimental data. Precise measurements are expressed in decimals to compare and analyze the results of experiments.

Temperature Variations: Temperature is an important quantity in science, and decimals are used to express temperature variations precisely, especially in fields like climatology and thermodynamics.

Decimals play a fundamental role in science by allowing for the precise expression of measurements, data, and experimental results. This is essential for advancing science and ensuring that scientific conclusions are supported by reliable data and precise calculations.

Finance: In the financial world, decimals are crucial for expressing amounts of money accurately. They are used in calculations of interest, rates, investments, debts, budgets, and other financial applications.

Decimals are vital in finance, where precision in calculations and the representation of monetary amounts is essential.

Expression of Monetary Amounts: In finance, monetary amounts are expressed in decimal form to reflect precision in transactions and financial calculations. For example, expressing $1,500.75 represents the exact amount of money.

Interest Calculations: Decimals are fundamental in calculating interest, whether in savings accounts, loans, or investments. Interest rates are expressed in decimal form (for example, 5% is expressed as 0.05) for accurate calculations.

Return Rates: In investments, decimals are used to express rates of return. This allows investors to calculate the return on their investments accurately and make informed decisions.

Budgets and Accounting: In preparing personal or business budgets, decimals are used to express expenses and income precisely. This is essential for determining financial status and the viability of a budget.

Division of Expenses and Profits: Decimals are used to distribute expenses and profits among different items in a company. This aids in financial management and cost analysis.

Amortization Calculations: In mortgage loans and other long-term loans, decimals are used to calculate amortization payments. This allows borrowers to understand how much they are paying and how much is applied to the principal.

Investment Analysis: In corporate finance, decimals are used to evaluate the profitability of investment projects. Future cash flows are discounted at accurate interest rates to determine the net present value (NPV) of an investment.

Dividends and Earnings per Share: In stock and company analysis, decimals are used to express earnings per share (EPS) and dividends per share. This is fundamental in evaluating investments in the stock market.

Foreign Exchange Markets: In the foreign exchange market, exchange rates are expressed in decimal form to reflect the precise rates at which currencies can be exchanged.

Precision in the representation of monetary amounts and financial calculations is essential for making informed decisions in the financial world. Decimals are the foundation of this precision and are an integral part of everyday financial operations.

Measurements and Unit Conversion: Decimals are used to express measurements of length, weight, volume, and other units. They are also essential in unit conversion, such as converting miles to kilometers or pounds to kilograms.

Decimals play a fundamental role in measurement and unit conversion across various disciplines as they allow for precise expression of quantities in different unit systems. Here are some ways decimals are used in measurements and unit conversions:

Length Measurements: In length measurements, such as the distance traveled on a trip or the dimensions of an object, decimals are used to express fractions of a unit of length (e.g., meters or feet). For example, 2.5 meters represents two and a half meters.

Weight Measurements: In weight measurements, decimals allow for expressing fractions of a unit of weight (such as grams or pounds). For example, 1.25 pounds represents one pound and a quarter.

Volume Measurements: In volume measurements, decimals are used to express fractions of a unit of volume (such as liters or gallons). For example, 0.75 liters represents three quarters of a liter.

Unit Conversions: Decimals are fundamental in unit conversions. For example, to convert 1 mile to kilometers, the conversion factor 1 mile = 1.60934 kilometers is used. This involves decimal use for accurate conversion.

Temperature: In temperature measurements, such as degrees Celsius or Fahrenheit, decimals are used to express temperature variations precisely. For example, a variation of 0.5 degrees Celsius represents half a degree.

Density: In chemistry and physics, density is expressed in units of mass per unit of volume. Decimals are essential for expressing density accurately.

Currency Conversion: In finance and international commerce, decimals are used to express exchange rates and perform currency conversions. This is fundamental in international trade and financial transactions.

Scientific Notation: In some scientific disciplines, such as physics and astronomy, decimals are used in scientific notation to represent extremely large or small values in a more compact and manageable form.

The ability to use decimals to express measurements and perform unit conversions accurately is essential in a variety of fields, from science and engineering to economics and everyday life. It facilitates communication and work with quantities that are not complete whole numbers, which is common in many applications.

Statistics: In statistics, decimals are fundamental for calculating averages, standard deviations, percentages, and other values of interest. They are also used in the representation of numerical data in graphs and tables.

Decimals play an essential role in statistics, where they are used to perform precise calculations and accurately represent numerical data. Here are some ways decimals are employed in statistics:

Average Calculation: Decimals are fundamental in calculating averages, such as the arithmetic mean. Decimal values allow for precise average expression, which is essential for summarizing data meaningfully.

Standard Deviations: The standard deviation is a measure of dispersion that indicates how spread out the data is around the average. Decimals are used to express the standard deviation accurately.

Percentages and Proportions: Decimals are used to express percentages and proportions. For example, if 20% of a population is in a particular group, this is expressed as 0.20 in decimal notation.

Data Representation in Graphs: In statistical graphs, such as histograms and bar charts, decimal values are used to represent numerical data accurately on the vertical axis. This allows for precise visualization of data distribution.

Regression Analysis: Decimal values are fundamental in regression analysis, where mathematical models are fitted to observed data. This involves working with precise values to find relationships between variables.

Probability and Inferential Statistics: In probability calculations and inferential statistics, decimals are used to calculate probabilities, confidence intervals, and z-scores, among other values. Precision is key in these calculations.

Sampling Statistics: In sampling studies, decimals are used to express estimates of population parameters based on samples. This is essential for providing accurate estimates.

Scientific Notation: In statistics, especially when dealing with large or small numbers, decimals are expressed in scientific notation to facilitate handling high-precision data.

Precision in statistics is crucial for informed decision-making and data analysis. Decimals enable accurate calculations and meaningful representation of numerical data, which is essential in scientific research, market studies, economic analyses, and many other statistical applications.

Technology and Programming: Decimals are used in programming to perform high-precision numerical calculations in scientific, financial, and engineering applications. They are also important in storing floating-point data.

Decimals are essential in technology and programming, allowing for high-precision calculations across a wide range of applications, from science and engineering to finance and computing. Here are some ways decimals are utilized in technology and programming:

Scientific Calculations: In scientific applications, precise calculations are fundamental. Decimals are used to represent measurements and experimental results with maximum accuracy.

Simulations and Models: In programming, decimals are used in simulations and mathematical models to represent continuous variables with high precision. This is important in fields like physics and computational biology.

Finance and Accounting Software: In financial and accounting software, decimals are crucial for performing precise financial calculations, including interest rates, investments, and detailed accounting.

Engineering and CAD Software: In engineering, decimals are used in computer-aided design (CAD) software to model three-dimensional objects and perform structural and engineering calculations.

Floating Point Data Storage: In programming, decimal numbers are stored in floating-point formats to allow for precise calculations in computer systems. These formats are essential in fields like graphics computing, artificial intelligence, and simulation.

Precision Calculations in Computational Mathematics: In computational mathematics, decimals are used to perform high-precision numerical calculations, such as calculating square roots or logarithms of non-integer numbers.

Signal Processing: In signal processing applications, such as digital music and image processing, decimals are important for performing mathematical operations and transformations with high precision.

Human-Computer Interaction: In user interface applications, such as mobile apps and websites, decimals are used to represent precise values, such as geographical coordinates or time measurements.

Financial Programming: In financial programming, decimals are essential for calculating interest rates, loan amortization, and financial projections.

In summary, decimals are an integral part of programming and technology, enabling precise calculations and high-accuracy data representation across a wide variety of applications. Precision is vital in these fields to ensure that results are reliable and useful.

Engineering: In engineering, decimals are used to measure and design components with precision. They are also crucial in performance calculations, structural analysis, electrical systems, and other technical applications.

Decimals play a critical role in engineering, allowing for precise measurements, accurate component design, and technical calculations for project development. Here are some ways decimals are used in engineering:

Precision Measurements: In engineering, making precise measurements of lengths, dimensions, and tolerances is essential. Decimals allow for high-accuracy measurement representation, which is fundamental in manufacturing and component design.

Design and CAD: In computer-aided design (CAD), decimals are used to model and represent components and systems with high precision. This is essential in the mechanical and civil engineering industries.

Structural Analysis: In structural engineering, decimals are used to calculate stresses, strains, and loads in structures like bridges and buildings. Precision in these calculations is essential to ensure safety and performance.

Electrical and Electronic Systems: In electrical and electronic engineering, decimals are fundamental in the design and analysis of circuits, control systems, and electrical components. They are used to represent resistances, currents, and voltages with high accuracy.

Fluid Mechanics: In fluid mechanics applications, such as aerodynamics and hydraulics, decimals are used to calculate flows, pressures, and velocities accurately.

Energy Efficiency: In energy efficiency projects, decimals are used to perform calculations on the performance of heating, cooling, and renewable energy systems. This is fundamental in energy and environmental engineering.

Process Optimization: In process and manufacturing engineering, decimals are used in calculations of performance, quality, and costs. This is vital for optimizing production and reducing waste.

Geotechnical and Surveying Applications: In geotechnical and surveying applications, decimals are used to represent elevations, coordinates, and land measurements with high precision. This is essential in civil and construction engineering.

Technical Data Analysis: In engineering, decimals are used to analyze technical data, such as laboratory test results, performance measurements, and quality data. Precision is crucial in making technical decisions.

Quality Control and Standards Compliance: Decimals are important in evaluating compliance with standards and technical specifications in engineering. This ensures that products and systems meet safety and quality requirements.

In summary, decimals are an essential part of engineering, allowing for precise measurements, technical calculations, and exact designs across a wide variety of technical applications. Precision is critical to ensure the performance and safety of engineering projects.

Medicine: In medicine, decimals are used to express measurements precisely, such as medication dosages, laboratory test results, and the dimensions of tissues and organs.

Decimals are a versatile and powerful tool for representing fractions and precise quantities across a wide variety of fields. They facilitate working with fractional values and enable accurate calculations and measurements, which is essential in many disciplines and practical applications.

8.Percentage: A way to express a fraction out of 100, like 25% is equal to ¼

A percentage is a way of expressing a fraction in terms of one hundred equal parts. It is used to represent the proportional relationship between a part and the whole. The word "percentage" comes from the Latin "per centum," which means "by a hundred."

The percentage symbol (%) is used to indicate that a quantity is expressed in relation to one hundred percent (100%). In other words, when we express something in terms of percentage, we are dividing that quantity into one hundred equal parts and representing how many parts of one hundred we have.

For example, if we say something is "20 percent," we mean it is equal to 20 parts out of every hundred possible parts. This is equivalent to the fraction 20/100, which can be simplified to 1/5. Thus, in percentage terms, 20 percent is equal to 1/5.

Percentages are a common way to communicate proportional relationships in a wide variety of contexts. They are used in situations ranging from calculating discounts and interest rates to reporting on population growth, approval ratings, tax rates, and many other quantitative data. Percentages are an essential tool for representing and understanding data in mathematics, economics, statistics, and other disciplines.

Percentages are a fundamental tool in mathematics and everyday life. Expressing fractions in terms of one hundred equal parts through percentages simplifies the understanding and comparison of proportional relationships in various situations.

Simplification of Proportional Relationships: By expressing fractions in terms of percentages, we are converting proportional relationships into a more intuitive form. Instead of dealing with more complicated fractions, such as 5/7 or 3/8, which may not be as visually clear, percentages represent these relationships on a scale of 0 to 100. This makes it easier for people to understand the magnitude of the relationship. For example, knowing something is 70 percent is more intuitive than saying it is 7/10.

The simplification of proportional relationships through percentages is a very practical and understandable approach. Here are more details on why this approach is beneficial:

Intuitiveness: Percentages are more intuitive for most people than fractions. The reason is that they express a direct relationship between a quantity and the whole, in terms of parts per hundred. This means a number in percentage is interpreted as a quantity in relation to a total. For example, 70 percent is easily understood as "70 out of every 100," which is more natural to visualize and comprehend than 7/10.

Direct Comparison: Percentages allow for a straightforward and simple comparison of magnitudes in relation to the whole. When looking at two numbers in percentage, it is evident which is greater or lesser in terms of the part they represent in one hundred. This is particularly useful in situations where decisions must be made based on numerical comparisons, such as choosing between two investment options with different rates of return.

Practical Applications: Percentages are used in everyday life to describe discounts, tax rates, interest rates, population growth, and many other proportional relationships. Most people are familiar with the concept of percentage and use it routinely in financial and business situations.

Facilitates Communication: When presenting data to diverse audiences, percentages are more effective for conveying information clearly and concisely. Graphs and tables that use percentages make it easier to communicate complex data to a non-specialized audience.

Percentages are a way to represent proportional relationships intuitively and accessibly for most people. They facilitate understanding and comparison of data across a variety of contexts, from economics and finance to statistics and personal decision-making. The simplicity and ease of interpretation make percentages a valuable tool in data communication and analysis.

Data Comparison: Percentages allow for easy comparison between different sets of data. When data is presented in percentage terms, it is simpler to compare and evaluate which of the data is larger or smaller in relation to the total. This is valuable in decision-making situations, such as comparing interest rates to choose a financial investment.

The ability to compare data efficiently and effectively is one of the key advantages of expressing information in percentage terms.

Standardization: Percentages represent all proportional relationships on a standard scale of 0 to 100. This means that regardless of the specific units or magnitudes involved, percentages allow data to be placed on the same scale, simplifying comparison.

Visual Comparison: Percentage numbers are easily comparable visually. When looking at two percentage values, it is simple to determine which of the two is greater or lesser in relation to the total. This is particularly useful when decisions must be made based on numerical comparisons, such as when assessing interest rates on different investment options.

Facilitates Decision-Making: In decision-making situations, such as financial investments or product choices, the ability to quickly compare rates of return, discounts, or interest rates in percentage terms helps individuals and businesses make more informed and efficient decisions.

Clarity in Reporting: When presenting data in reports, graphs, or tables using percentages, the clarity of the information is improved. This facilitates communication of results and trends to diverse audiences, even to those who do not have a deep understanding of the units or magnitudes involved.

Comparative Analysis: Percentages are essential in comparative data analysis in fields like economics and social sciences. They allow for effective evaluation and communication of changes in populations, growth rates, inflation rates, unemployment rates, and other economic and social indicators.

Risk and Return Assessment: In finance, percentages are crucial for assessing the risk and return on investments. Investors can easily compare return rates, asset yields, and interest rates across different financial instruments.

Percentages provide a uniform and comprehensible framework that facilitates data comparison across a wide variety of contexts. This not only enhances informed decision-making but also contributes to the efficient communication of information in reports, graphs, and presentations. The ability to compare data in percentage terms is a valuable skill in business, economics, statistics, and many other disciplines.

Applications in Finance: Percentages are essential in finance and economics. They are used to calculate taxes, interest rates, discounts, and profits. Additionally, they help individuals evaluate investments, loans, and other financial decisions.

Percentages are a fundamental tool in finance and economics, applied in various contexts to make informed decisions and perform accurate calculations.

Tax Calculation: Percentages are used to calculate income, sales, property, and other types of taxes. For example, the government may set a tax rate of 20 percent on income, meaning that 20 percent of total income must be paid in taxes.

Interest Rates: Percentages are essential in evaluating loans and interest rates. When applying for a loan, the interest rate is expressed as an annual percentage (for example, a rate of 5 percent per year). This helps borrowers understand how much they will pay in interest.

Discounts and Offers: In retail, percentages are used to calculate discounts and offers. For example, a retailer may offer a 10 percent discount on a product. Consumers can easily evaluate how much they save.

Investment Evaluation: Investors use percentages to assess the performance of their investments. They can calculate the percentage return on investment (ROI) to understand how much they have gained or lost on their investments.

Profit and Margin Evaluation: Businesses use percentages to evaluate their profitability. The profit margin is calculated as a percentage of total sales and shows how much remains as profit after covering costs.

Risk Analysis: Percentages are also valuable in financial risk analysis. Default rates and interest rates are expressed in percentage terms, helping financial institutions assess and manage credit risk.

Investment Comparison: Investors can compare different investment opportunities using percentage return rates. This helps them determine which option is the most profitable.

Personal Budgeting and Expenses: Individuals use percentages to manage their personal finances. They can calculate how much they spend on different categories relative to their total income.

Loan Assessment: When applying for a mortgage, for instance, borrowers evaluate the interest rate expressed in percentage to understand the total cost of the loan over time.

Inflation Estimation: Economists and financial analysts use inflation rates expressed in percentage to estimate how prices of goods and services are increasing.

Percentages are an essential tool in finance and economics. They facilitate calculation, comparison, and informed financial decision-making for both individuals and businesses. Understanding how percentages are applied in these contexts is crucial for making sound financial choices and effectively managing economic resources.

Measuring Growth and Change: Percentages are used to measure growth and change in various areas, from a city's population to the price increase of a product. This provides a clear understanding of the magnitude of changes over time.

Measuring growth and change is a fundamental aspect across a wide variety of fields, and percentages are an essential tool for quantifying and understanding these changes effectively.

Population: In demography, percentages are used to measure population growth. For example, a 2 percent increase in the population of a city indicates that the population has grown by 2 percent compared to the previous total.

Economic Growth: Percentages are crucial in measuring economic growth. A country's Gross Domestic Product (GDP) can be expressed in terms of percentage growth, showing how the economy has increased or decreased over a given period.

Inflation: The inflation rate is expressed as a percentage and is used to measure the average increase in prices of goods and services. Percentages help individuals and businesses understand the impact of inflation on their purchasing power.

Interest Rates: Changes in interest rates are expressed in percentage terms and have a significant impact on the economy. Higher or lower interest rates affect the cost of loans and investments.

Revenue Growth: Percentages are used to measure the growth of personal or business income over time. A 5 percent increase in income means that income has grown by 5 percent compared to the previous period.

Price Growth: In economics, percentages are applied to measure price increases of goods and services, which is fundamental for assessing the economic impact on consumers and businesses.

Market Growth: Companies use percentages to measure market growth. For example, if a market has grown by 10 percent, this indicates higher demand for products or services in that sector.

Investment Growth: Investors use percentages to assess the growth of their investments. An 8 percent return on an investment portfolio indicates how much the portfolio's value has increased.

Sales Growth: In sales and marketing, percentages are used to evaluate the growth of product or service sales. A 15 percent increase in sales indicates significant growth in total sales.

Employment Growth: Percentages are applied to measure employment growth in a region or industry. A 3 percent increase in employment indicates growth in the workforce.

Percentages are essential for measuring growth and change across a wide range of fields, from demographics and economics to finance and marketing. They provide a standardized and comprehensible way to quantify and communicate the magnitude of changes over time, which is fundamental for decision-making and trend evaluation.

Data Presentation: Percentages are an effective way to present data in reports, graphs, and presentations. Bar charts or pie charts with percentages facilitate the communication of complex information clearly and visually.

Data presentation is a crucial aspect of communicating information and findings across a wide variety of contexts, from research reports to business presentations and education. Percentages are a powerful tool for effectively and understandably representing data.

Bar Charts: Bar charts are a common way to present data in percentage terms. For example, a bar chart can show the percentage distribution of expenses in a household budget, making it easier to visualize how financial resources are allocated.

Pie Charts: Pie charts divide a circle into sectors that represent percentages. These charts are effective for showing how a total quantity is divided into several parts. For example, a pie chart can represent the distribution of sales by product category.

Line Charts: Line charts are used to show trends over time. Percentages can be a useful way to represent changes over time, such as GDP growth, inflation rates, or market trends.

Tables and Charts: In reports and presentations, tables and charts often include percentages to summarize data and show comparisons. For example, a table may show the approval ratings of different products based on consumer surveys.

Infographics: Infographics effectively use percentages to communicate complex information visually and attractively. For example, an infographic may show the percentage distribution of diseases in a population.

Comparisons and Analysis: By using percentages in data presentation, it becomes easier to compare and analyze information. Readers or audiences can quickly and clearly identify which of the categories or elements is the largest or smallest in relation to the total.

Research Reports: In fields such as science, health, and economics, percentages are common in research reports. They are used to communicate findings, survey results, and data analysis in a way that is easy to understand.

Business Presentations: In the business environment, percentages are used in presentations to highlight key metrics, financial achievements, and sales results. This helps teams and investors understand the company's performance.

Education: Percentages are taught in mathematics and are used in education to assess student progress. They are also applied in data presentation in lessons and teaching materials.

Percentages are an effective tool for presenting data clearly and visually. They facilitate the communication of complex information and help recipients understand and analyze data more efficiently. Graphs and tables that use percentages are common in a wide variety of contexts, from business decision-making to presenting academic research results.

Precise Calculation Tool: The percentage formula, which relates the part to the whole in terms of one hundred parts, is a useful mathematical tool for precise calculation. It is especially useful for calculating discounts, taxes, growth rates, and other proportional relationships in a wide variety of contexts.

The percentage formula is indeed a powerful and versatile mathematical tool that allows for precise calculation of the relationship between a part and the whole in terms of one hundred parts. This formula is fundamental in a variety of contexts and applications.

Percentage Formula: The percentage is calculated using the following formula: Percentage = (Part / Total) × 100

The "Part" represents the quantity being considered in relation to the whole or total. The "Total" is the complete amount or set to which the part belongs. Multiplying by 100 converts the fraction into a percentage.

Utility of the Percentage Formula:

Discounts and Price Increases: The percentage formula is commonly used in sales and commerce to calculate discounts (for example, a 20% discount on an item) or price increases (for example, a 10% increase in the cost of a product).

Taxes and Rates: In tax and financial contexts, it is used to calculate taxes (percentage of income) and rates (percentage interest on loans or investments).

Growth and Decline: Percentages are used to calculate the growth or decline of values in fields such as economics (GDP growth), finance (investment returns), and demographics (population growth).

Probabilities and Statistics: In statistics, percentages are used to express probabilities and frequencies of events, which is fundamental in data analysis and probability studies.

Comparisons and Evaluations: Percentages facilitate comparison and evaluation of proportional relationships. For example, comparing approval ratings of two political candidates.

Data Presentation: As mentioned earlier, percentages are an effective way to present data in reports, graphs, and presentations, making the information more accessible and understandable.

Decision-Making: Percentages help individuals and organizations make informed decisions. For example, calculating the percentage of sales growth can influence strategic marketing decisions.

Resource Management: In management and planning, percentages are essential for managing resources, such as budgets and workforce.

Business Data Analysis: In the business world, percentages are used to evaluate key metrics such as profit margins, conversion rates, and inventory turnover rates.

The percentage formula is a powerful mathematical tool that applies in a wide variety of situations in everyday life, business, finance, statistics, and other fields.

9.Geometry: The study of shapes, sizes, and properties of objects and figures

Geometry is a branch of mathematics focused on the study of shapes, sizes, and properties of objects and figures in space. Mathematicians and geometers analyze how figures relate to one another, how they can be transformed, and how they can be measured. Geometry is divided into several branches, which include:

Euclidean Geometry: This is the classical geometry developed by the ancient Greek mathematician Euclid. It is based on Euclid's axioms and focuses on the properties of points, lines, planes, and three-dimensional figures in space. Euclidean Geometry, also known as classical geometry, is a geometric system developed by the ancient Greek mathematician Euclid, who lived around 300 BC. Euclid wrote an influential treatise called "The Elements," which established the foundations of Euclidean geometry. This form of geometry is based on a series of axioms or postulates, which are basic statements considered true without the need for proof. Some of the most well-known postulates include:

The line postulate: Through any two distinct points, there passes a unique line.

The segment postulate: A straight line segment can be extended indefinitely.

The angle postulate: Given two angles and a point, a third angle can be drawn that is equal to one of the two given angles and adjacent to the other.

The parallel postulate: Through a point outside a line, only one line can be drawn parallel to that line.

Euclidean geometry concentrates on the study of the properties of points, lines, planes, and three-dimensional figures in space, using these axioms as a foundation. Additionally, it develops using a series of definitions and theorems that are derived from these axioms. It is a fundamental geometric system in mathematics and has been widely used in teaching geometry due to its clarity and simplicity.

However, throughout the history of mathematics, it was discovered that Euclidean postulates were not the only possible ones, and other geometries were developed, such as hyperbolic geometry and elliptic geometry, which present alternative postulates and have expanded our understanding of geometric structures. These new geometries are known as non-Euclidean geometries and are fundamental in modern physics, especially in Einstein's theory of relativity.

Analytic Geometry: This branch combines geometry with algebra, allowing geometric figures to be described using mathematical equations. This facilitates the study of geometry in a coordinate system. Analytic Geometry is a branch of mathematics that combines geometric concepts with algebraic techniques. This discipline allows for the description of geometric figures using mathematical equations and facilitates the study of geometry in a coordinate system. Some of the most important aspects of analytic geometry include:

Coordinate System: In analytic geometry, a coordinate system is used to represent points, lines, and figures in either the plane or space. The most common coordinate systems are the Cartesian coordinate system in two dimensions (plane) and the three-dimensional coordinate system in three dimensions (space).

Points and Vectors: Points in space are represented as tuples of numbers (coordinates) that indicate their position in the coordinate system. Vectors are used to represent displacements between points and are expressed through magnitudes and directions.

Equations of Lines and Curves: Lines and curves are represented by algebraic equations. For example, the equation of a line in a 2D Cartesian coordinate system is of the form $y = mx + b$, where m is the slope and b is the y-intercept. This allows for the description of straight lines in algebraic terms.

Distances and Angles: Analytic geometry facilitates the calculation of distances between points and angles between lines or vectors using algebraic techniques and mathematical formulas.

Geometric Transformations: Analytic geometry is also used to study geometric transformations, such as translations, rotations, and scalings, in algebraic terms. These transformations are expressed using matrices and algebraic operations.

Analytic geometry is a powerful tool that has found applications in various fields, including physics, engineering, computer graphics, computational geometry, and other disciplines. It allows for geometric problems to be solved more efficiently and accurately by translating them into mathematical terms, facilitating the analysis and resolution of problems in geometry.

Differential Geometry: This branch focuses on the study of curves and surfaces and relates to concepts such as tangents, curvature, and other differential properties of geometric figures. Differential Geometry is a branch of mathematics that focuses on the study of curves and surfaces in space and is closely related to differential concepts to analyze geometric properties. Some key concepts in Differential Geometry include:

Curves and Surfaces: Differential Geometry concentrates on the study of curves in the plane and surfaces in three-dimensional space. These can be simple curves, such as straight lines or circles, or more complex curves, such as particle trajectories or surface boundaries.

Tangent Vectors: In Differential Geometry, tangent vectors are used to describe the direction and rate of change at a point on a curve or surface. These tangent vectors are fundamental for understanding the geometry of the curve or surface at that point.

Curvature: Curvature is a central concept in Differential Geometry. It indicates how "curved" or "twisted" a curve or surface is at a given point. Curvature relates to the rate of change of the tangent vector and is used to measure the "inclination" of the curve at that point.

Riemannian Geometry: This is an extension of Differential Geometry that focuses on the geometry of curved surfaces and metric spaces. It is used to study geometry in spaces where geometric properties can vary from point to point.

Fundamental Theorem of Differential Geometry: This theorem establishes an important relationship between the curvature of a closed curve and the number of times the curve encloses a point in space.

Differential Geometry is fundamental in physics, especially in Albert Einstein's theory of relativity, where it is used to describe the curvature of space-time. It also applies in cartography, robotics, surface modeling in computer graphics, and many other areas where understanding and analyzing geometric properties of curves and surfaces is important.

Projective Geometry: This branch deals with the properties of figures that do not change under projective transformations, meaning it focuses on properties that are independent of perspective. Projective Geometry is a branch of geometry that focuses on the properties of geometric figures that remain unchanged under projective transformations. In other words, it focuses on properties that are independent of perspective or how a projection is made. Some fundamental concepts of Projective Geometry include:

Projective Transformations: Projective transformations are those that preserve the projective properties of geometric figures. These transformations include projections from a point or a perspective projection, as well as affine transformations like translations, rotations, and scalings.

Points, Lines, and Incidence: In Projective Geometry, there is no absolute distance between points or an absolute notion of parallelism. Instead, the notion of incidence is used, which refers to the relationship between points and lines. The intersection of a line with a point is called incidence.

Desargues' Theorem: This theorem is one of the fundamental results in Projective Geometry and relates to the property of homology. It states that if two triangles are in perspective from a point, then the vertices of the triangles lie on a line.

Perspective Projectivity: Perspective projectivity is a key concept in Projective Geometry. It indicates that two projective figures can be considered equivalent if there is a projective transformation that relates them.

Applications in Art and Computer Graphics: Projective Geometry is used in artistic applications, such as painting and photography, to represent three-dimensional objects on a two-dimensional surface. Additionally, in computer graphics, it is applied to create illusions of perspective and simulate how objects appear from different viewpoints.

Projective Geometry is a branch of mathematics with applications in architecture, art, design, and technical fields such as cartography, computer vision, and algebraic geometry. Its focus on properties that are invariant under projective transformations makes it especially useful for the study of perspectives and the representation of objects in various contexts.

Non-Euclidean Geometry: This studies geometries in which Euclid's postulates do not hold, such as hyperbolic geometry and elliptic geometry, which have applications in relativity theory and topology, respectively. Non-Euclidean Geometry refers to the study of geometries in which Euclid's postulates do not hold. Unlike Euclidean Geometry, which is based on Euclid's postulates and develops in a flat or three-dimensional space with well-defined properties, Non-Euclidean Geometry considers other possibilities. The two most well-known forms of Non-Euclidean Geometry are Hyperbolic Geometry and Elliptic Geometry:

Hyperbolic Geometry: In Hyperbolic Geometry, Euclid's fifth postulate, the parallel postulate, is rejected. This leads to a space where, given a line and a point outside it, there are multiple parallel lines that pass through the point and do not intersect the original line. This results in geometric properties that are very different from those of Euclidean Geometry. Hyperbolic

Geometry has applications in relativity theory, particularly in describing the geometry of space-time in the presence of strong gravitational fields.

Elliptic Geometry: In Elliptic Geometry, several of Euclid's postulates are rejected, including the fifth postulate regarding parallels. In this type of geometry, no parallel lines exist at all. All lines intersect at some point. Elliptic Geometry has properties that are quite different from both Euclidean and Hyperbolic Geometry. It is used in topology, number theory, and in representing the surface of a sphere.

Both Non-Euclidean Geometries have important applications in mathematics and physics. Hyperbolic Geometry is fundamental to Einstein's general theory of relativity, which describes gravity in terms of the curvature of space-time. Elliptic Geometry is significant in areas of mathematics such as number theory and topology, and it is also used in cartography to accurately represent the surface of the Earth. These non-Euclidean geometries have expanded our understanding of geometric structures and have led to significant advances in mathematical theory and modern physics.

Fractal Geometry: This deals with highly irregular and self-similar figures that repeat at different scales. Fractals are used in various fields, such as chaos theory and the representation of natural structures. Fractal Geometry is a branch of geometry that focuses on the study and description of highly irregular and self-similar figures that repeat at different scales. Fractal objects exhibit patterns and complex structures that often resemble themselves when observed at different levels of magnification. Some key concepts of Fractal Geometry include:

Self-Similarity: Self-similarity is a fundamental characteristic of fractals. It means that a portion of the fractal resembles a miniature or scaled version of the entire fractal. This property repeats at different levels of magnification.

Fractal Dimension: Unlike traditional geometric objects, such as straight lines (dimension 1) or flat surfaces (dimension 2), fractals have non-integer or fractional fractal dimensions. This reflects their complexity and the way they fill space unconventionally.

Iteration: Fractals are often generated through iterative processes. A set of simple rules is repeatedly applied to build the complete fractal, resulting in increasingly detailed patterns.

Chaos Theory: Fractal Geometry has connections with chaos theory. Some chaotic dynamic systems can be described using fractals, and fractals can be used to represent complex and turbulent systems.

Applications: Fractals are used in various fields, including modeling nature and natural phenomena such as irregular coastlines, tree structures, climate systems, and cellular structures. They are also applied in image compression, landscape generation in computer graphics, and information theory.

One of the most well-known fractals is the Mandelbrot Set, which is a highly complex and self-similar figure generated through simple mathematical iterations. Fractal Geometry has had a significant impact on understanding complexity in nature and representing structures and phenomena that do not conform to traditional geometric models.

Geometry has applications in many areas of science and technology, including physics, architecture, engineering, cartography, computer graphics, and computational geometry. It is also fundamental in solving mathematical problems and understanding abstract concepts in mathematics. Geometry plays a fundamental role in many areas of science, technology, and mathematics.

Physics: Geometry is fundamental in physics, especially in mechanics, optics, and the theory of relativity. It is used to describe and analyze the position, movement, and properties of objects and particles in space and time.

Architecture and Engineering: In architecture and civil engineering, geometry is essential for designing and constructing structures, buildings, and bridges. Geometric principles are used to ensure the stability and safety of constructions.

Cartography: Cartography relies on geometry to represent the Earth on maps. Cartographic projections and geometric techniques are used to accurately map geographical areas onto flat surfaces.

Computer Graphics: In the computer graphics industry, geometry is essential for creating and rendering 2D and 3D images. Geometric concepts are used to model three-dimensional objects and simulate realistic visual effects.

Computational Geometry: This branch of computer science focuses on developing algorithms and techniques to solve geometric problems, such as line intersections, area and volume calculations, and representing objects in three-dimensional spaces.

Topology: Topology is a branch of mathematics based on geometry to study spatial properties and relationships between shapes, but without concern for distances and precise measurements. It has applications in knot theory, graph theory, and algebraic geometry.

Mathematical Problem Solving: Geometry is a valuable tool for solving mathematical problems and proving theorems in various areas of mathematics. It helps visualize and understand abstract concepts.

Biology and Natural Sciences: Geometry is used to model and analyze biological structures and natural processes. For example, in molecular biology, geometric concepts are used to study the structure of molecules.

Robotics: In robotics, geometry is applied to control and plan the movement of robots and understand the kinematics of mechanical systems.

Economics and Social Sciences: Geometry is sometimes used in mathematical models and statistics in fields such as economics and geography to analyze data and spatial relationships.

Geometry is a versatile discipline that plays an essential role in a wide range of applications, from the exact sciences to the social sciences, making it one of the most fundamental branches of mathematics.

10.Triangles: Figures with three sides and three angles

A triangle is a flat geometric figure that has three sides and three angles. Triangles are one of the most basic and fundamental shapes in geometry.

Three Sides: As its name indicates, a triangle has three sides. The sides are commonly referred to as "a," "b," and "c." Triangles are fundamental in geometry and are characterized by having three sides. These sides are usually labeled as "a," "b," and "c," and are used to describe the length of each of the triangle's sides. The relationships between the lengths of the sides of a triangle are essential for the study of trigonometry and geometry, and they are used in a wide variety of mathematical and scientific applications.

The relationships between the lengths of the sides of a triangle are fundamental in mathematics, particularly in trigonometry and geometry, and have a wide range of applications in sciences and related disciplines. Some key points about these relationships are:

Pythagorean Theorem: This theorem establishes an important relationship in right triangles. It states that in a right triangle, the square of the length of the hypotenuse (the side opposite the right angle) is equal to the sum of the squares of the lengths of the other two sides. It is fundamental in trigonometry and has applications in geometry and physics.

The Pythagorean Theorem is one of the most fundamental results in geometry and trigonometry. This theorem specifically applies to right triangles, which are triangles that have a 90-degree angle (a right angle). The theorem states the following:

In a right triangle, the sum of the squares of the lengths of the two legs (the two sides that form the right angle) is equal to the square of the length of the hypotenuse (the side opposite the right angle).

This theorem is essential in trigonometry and is used to solve a variety of problems related to right triangles. Some of its applications include:

Calculating Lengths: You can use the Pythagorean Theorem to find the length of an unknown side in a right triangle if you know the lengths of the other two sides.

Trigonometry: The theorem is fundamental in trigonometry, as it allows you to define and relate trigonometric functions such as sine, cosine, and tangent in terms of the lengths of the sides of a right triangle.

Geometry and Construction: It is used in construction and design to ensure right angles and calculate distances.

Navigation: The Pythagorean Theorem is essential in navigation for determining distances on maps and nautical charts.

Physics: It is applied in physics, especially in kinematics and in problems related to the motion of particles.

Trigonometry: Trigonometry is a branch of mathematics that focuses on the relationships between the angles and sides of a triangle. Concepts such as sine, cosine, and tangent are used to relate angles and side lengths in triangles.

Trigonometry is indeed an important branch of mathematics that focuses on studying the relationships between the angles and sides of a triangle. Through concepts like sine, cosine, and tangent, trigonometry allows relating angles and side lengths in triangles and extends to various applications in mathematics, science, and technology. Some key points include:

Trigonometric Functions: Trigonometric functions, such as sine (sin), cosine (cos), and tangent (tan), are used to describe the relationships between angles and sides in triangles. These functions are fundamental in trigonometry and are used to model periodic phenomena, solve geometry problems, and much more.

Triangles: Trigonometry mainly applies to triangles, focusing on right triangles (triangles that contain a 90-degree angle). These triangles are fundamental in trigonometry and are the basis for defining trigonometric functions.

Problem Solving: Trigonometry is used to solve a wide variety of problems in physics, engineering, navigation, astronomy, and many other disciplines. For example, it can be used to calculate distances, angles, velocities, and other physical quantities.

Graphical Representation: Trigonometric functions are graphically represented as sinusoidal waves, which is fundamental in signal analysis and periodic phenomena in sciences like physics and engineering.

Trigonometric Identities: Trigonometric identities are equations that relate trigonometric functions to one another. These identities are useful for simplifying expressions and solving trigonometric equations.

Trigonometry is a powerful and versatile tool in mathematics and in practical applications across a wide range of disciplines. Whether to calculate astronomical distances, analyze electrical signals, model harmonic motion, or solve problems in geometry, trigonometry is essential in modern science and technology.

Congruence and Similarity Theorems: In geometry, theorems are used to establish conditions under which triangles are congruent (having the same sides and angles) or similar (having equal angles and proportional sides).

Theorems of congruence and similarity are fundamental concepts in geometry and are used to compare and relate triangles. Here's a more detailed explanation of these concepts:

Congruence of Triangles: The congruence of triangles refers to the situation where two triangles have the same sides and angles, meaning they are identical in shape and size. For two triangles to be congruent, certain conditions or criteria must be met:

Side-Side-Side (SSS): If the three sides of one triangle are congruent to the three sides of another triangle, then the two triangles are congruent.

Side-Angle-Side (SAS): If two sides and the angle included between them of one triangle are congruent to two sides and the corresponding angle of another triangle, then the two triangles are congruent.

Angle-Side-Angle (ASA): If two angles and the side between them of one triangle are congruent to two angles and the corresponding side of another triangle, then the two triangles are congruent.

Hypotenuse and an Acute Angle (HA): If the hypotenuse and an acute angle of a right triangle are congruent to the hypotenuse and the corresponding acute angle of another right triangle, then the two triangles are congruent.

Similarity of Triangles: The similarity of triangles refers to the situation where two triangles have equal angles and proportional sides, meaning they are similar but not necessarily identical in size. The criteria for the similarity of triangles are as follows:

Angle-Angle (AA): If two triangles have two equal angles, then the third angles are also equal, and the two triangles are similar.

Side-Angle-Side (SAS): If two triangles have one proportional side, one angle, and another proportional side in the same order, then they are similar.

Side-Side (SS): If two triangles have their corresponding sides proportional, then they are similar.

The congruence and similarity of triangles are fundamental concepts in geometry and are used to solve problems related to geometry and trigonometry, as well as in practical applications in construction, surveying, navigation, and other fields. These theorems provide powerful tools for analysis and problem-solving in geometric contexts.

Solving Practical Problems: The relationships in triangles are applied in practical problems, such as navigation, construction, surveying, physics, engineering, and architecture. For example, understanding and applying trigonometric relationships is essential for determining distances or angles.

The relationships in triangles and trigonometric concepts are widely applied in solving practical problems across various fields. Here are examples of how they are used in everyday situations and specific disciplines:

Navigation: In maritime and aerial navigation, determining distances and precise directions is crucial. Trigonometric relationships are used to calculate distances between points, determine positions on a map, and navigate based on headings and angles.

Construction and Engineering: In construction and civil engineering, relationships in triangles and trigonometric properties are applied to measure and design structures, calculate loads and stresses, and determine angles of inclination, among other things.

Surveying: Surveyors use trigonometry to perform topographic surveys and map the Earth's surface. This involves accurately measuring distances and angles to create detailed maps.

Physics: In physics, trigonometric relationships are used to analyze the motion of objects, calculate forces, and solve problems related to kinematics and dynamics.

Architecture: In architecture, trigonometric properties are applied in the design and construction of buildings, especially in determining angles and proportions to ensure stability and aesthetics in structures.

Geology: Geologists use trigonometry to measure rock strata, angles of inclination, and other field data to understand the geology of a region.

Astronomy: Astronomers use trigonometric relationships to calculate the position and distance of celestial objects, as well as to make accurate astronomical observations and measurements.

Statistics: Trigonometry is used in statistics and data analysis to decompose and analyze periodic components in datasets.

These are just a few examples of how relationships in triangles and trigonometry are essential in solving practical problems across a wide range of fields. These mathematical concepts provide fundamental tools for measuring, calculating, and designing in disciplines ranging from navigation to construction and scientific research.

Modeling and Simulation: In fields such as data science and numerical simulation, relationships in triangles are used to model and solve problems in various domains, from economics to biology.

Modeling and simulation are important applications of relationships in triangles and trigonometric concepts in fields that extend beyond pure geometry and trigonometry. Here's a more detailed explanation of how they are used in these contexts:

Data Science: In data science, trigonometric relationships can be used to analyze data that exhibit periodic patterns. This may include time series analysis, where trigonometric techniques are applied to decompose signals into sinusoidal components and model seasonal trends.

Numerical Simulation: Numerical simulation is used to model systems and phenomena across various fields, from physics to engineering and biology. Relationships in triangles and trigonometric functions are often used to simulate the behavior of physical and biological systems, such as electrical circuits, planetary motions, and fluid dynamics.

Economics: In economics, trigonometric techniques and time series analysis are applied to model and predict economic patterns, such as business cycles and fluctuations in financial markets.

Biology: Biology employs mathematical models to understand and predict the behavior of biological systems. In the study of biological rhythms, such as circadian cycles, trigonometric relationships are useful for describing biological oscillations.

Geophysics: Geophysics focuses on studying the geological and physical processes of the Earth. Trigonometry is used to model and analyze the propagation of seismic waves and other phenomena related to geology.

In summary, relationships in triangles and trigonometric functions are versatile mathematical tools that are applied in a wide range of fields. Modeling and simulation are examples of how these tools are used to understand and predict phenomena across various scientific disciplines and practical applications. The ability to model complex systems and solve problems using these mathematical tools is a fundamental part of research and analysis in data science and numerical simulation.

Computational Geometry: Relationships between the sides and angles of triangles are used in algorithms and computational techniques to solve geometric problems, such as triangulating points in a plane.

Computational geometry is a branch of computer science and geometry that focuses on developing algorithms and techniques to solve geometric problems using geometric concepts and relationships between the sides and angles of geometric figures, including triangles. Here's more information on how relationships in triangles are applied in computational geometry:

Point Triangulation: One of the classic problems in computational geometry is triangulating points in a plane. Given a set of points in a plane, the goal is to divide the plane into triangles such that the points are vertices of those triangles. Trigonometric relationships are used to calculate the angles between points and determine the arrangement of triangles.

Intersection of Segments and Polygons: Computational geometry is applied in detecting intersections between line segments and polygons. This is useful in applications such as computer graphics, gaming, digital mapping, and computer-aided design.

Calculating Areas and Perimeters: Algorithms in computational geometry use the relationships between the sides and angles of geometric figures to calculate areas, perimeters, and other properties. For example, the areas of polygons and shaded areas in graphic applications can be calculated.

3D Visualization: In 3D visualization applications, geometric relationships are used to represent and transform three-dimensional objects, such as 3D models of buildings, terrains, and objects.

Robotics: Motion planning and kinematics of robots are based on geometric and trigonometric principles to calculate trajectories and positions of robot arms or moving parts.

Spatial Analysis: Computational geometry is used in spatial analysis in geographic information systems (GIS) to solve problems related to location, distance, and spatial relationships between geographic objects.

In summary, computational geometry plays an important role in solving problems that involve geometry and geometric relationships using algorithms and computational techniques. These algorithms have applications across a wide range of fields, from computer graphics to robotics and digital mapping.

Relationships in triangles and geometric properties are fundamental in creating efficient algorithms to solve these geometric problems.

Relationships in triangles are a key component of mathematics and have broad and varied applications in science, engineering, and many other disciplines. These relationships allow for solving practical problems and understanding abstract concepts in the natural and mathematical world.

Theorems and properties of triangles are a fundamental part of mathematics and apply in fields ranging from physics to architecture and engineering.

A triangle is a geometric figure consisting of three line segments that meet at their endpoints. Each of these line segments is called a "side" of the triangle. The fundamental characteristic of a triangle is that it always has three sides. The sides of a triangle can be identified and commonly labeled as "a," "b," and "c." These are simply names used to distinguish the sides of the triangle and allow us to refer to them clearly.

When working with triangles, it is common to use these labels to describe their properties and relationships. For example, if you are calculating the length of the sides of a triangle or applying trigonometric theorems, these labels will be used to specify which side you are measuring or which relationship you are calculating.

In summary, "a," "b," and "c" are common labels used to refer to the three sides of a triangle, and they are an essential part of the notation used in geometry to work with triangles and describe their properties.

Three Angles: A triangle also has three angles, which are usually denoted as angle A, angle B, and angle C.

Sum of Angles: The sum of the three angles of a triangle is always equal to 180 degrees. This property is known as the "sum of the internal angles of a triangle."

The property of the sum of the internal angles of a triangle is one of the fundamental concepts in geometry and applies to all triangles. This property states that the sum of the three interior angles of a triangle is always equal to 180 degrees (or π radians).

This property is independent of the type of triangle you are considering: whether it is an equilateral triangle (where all three sides and three angles are equal), an isosceles triangle (where at least two sides and two angles are equal), or a scalene triangle (where all sides and angles are different), the sum of the internal angles will always be equal to 180 degrees.

This property is essential in geometry and serves as a basis for many theorems and geometric proofs. It also helps to understand the relationship between the interior angles of a triangle and is applied in numerous geometric and trigonometric problems.

Classification by Sides: Triangles can be classified according to the lengths of their sides. For example, an equilateral triangle has all its sides of equal length, while a scalene triangle has all sides of different lengths.

Equilateral Triangle: An equilateral triangle is a type of triangle in which all sides have the same length. This means that all three sides are equal in length. Furthermore, because all sides are equal, all interior angles are also equal, and each interior angle measures 60 degrees. Equilateral triangles are a special case of isosceles triangles and are especially symmetrical.

Isosceles Triangle: An isosceles triangle has at least two sides of equal length. This means that two of the three sides are equal, while the third side is of different length. In an isosceles triangle, the angles opposite to the equal sides are also equal to each other.

Scalene Triangle: A scalene triangle is one in which all three sides have different lengths. No side is equal to another in length. Because the sides are of different lengths, the interior angles will also have different measures. A scalene triangle is the most general type of triangle.

These classifications are based on the lengths of the sides of a triangle and are fundamental in geometry. Each type of triangle has specific properties and characteristics that are used in the study of geometry and in solving problems related to triangles in mathematics and practical applications in various disciplines.

Classification by Angles: Triangles can also be classified according to the measure of their angles. For example, a right triangle has a right angle (90 degrees), while an obtuse triangle has an angle greater than 90 degrees.

Triangles can also be classified based on the measure of their angles, and this classification is another fundamental part of geometry.

Right Triangle: A right triangle is a triangle that has a right angle, that is, an angle of 90 degrees. The side opposite the right angle is called the hypotenuse, and the other two sides are called the legs. Right triangles are important in trigonometry and are used in a variety of applications, such as distance calculations, elevations, and projections.

Acute Triangle: An acute triangle is a triangle in which all interior angles are acute, meaning they measure less than 90 degrees. In an acute triangle, all three angles are acute, indicating that the sides are "pulled inward."

Obtuse Triangle: An obtuse triangle is a triangle that has at least one obtuse angle, meaning an angle that measures more than 90 degrees. The other two angles are acute. In an obtuse triangle, one or more sides are "opened outward" due to the obtuse angles.

These classifications are based on the measure of the interior angles of a triangle and are fundamental in geometry and trigonometry. Each type of triangle has specific properties and characteristics that are used in the study of geometry and in solving problems related to triangles in mathematics and practical applications in various disciplines.

Pythagorean Theorem: In a right triangle, the Pythagorean theorem states that the square of the length of the hypotenuse (the side opposite the right angle) is equal to the sum of the squares of the lengths of the other two sides.

The Pythagorean Theorem is one of the most fundamental results in geometry and specifically applies to right triangles, which are triangles that have a 90-degree angle (a right angle). This theorem states the following:

In a right triangle, the sum of the squares of the lengths of the two legs (the two sides that form the right angle) is equal to the square of the length of the hypotenuse (the side opposite the right angle).

This theorem is fundamental in trigonometry and is used in a variety of applications in geometry, mathematics, and physics.

Calculating Lengths: You can use the Pythagorean Theorem to find the length of an unknown side in a right triangle if you know the lengths of the other two sides.

Trigonometry: The theorem is fundamental in trigonometry as it allows defining and relating trigonometric functions such as sine, cosine, and tangent in terms of the lengths of the sides of a right triangle.

Geometry and Construction: It is used in construction and design to ensure right angles and calculate distances.

Navigation: The Pythagorean Theorem is essential in navigation for determining distances on maps and nautical charts.

Physics: It is applied in physics, especially in kinematics and in problems related to the motion of particles.

In summary, the Pythagorean Theorem is an essential mathematical tool used in numerous practical and theoretical applications, and it is an important foundation in the study of geometry and trigonometry.

Triangles are the basis of many areas of geometry and have applications in various disciplines, from trigonometry and analytic geometry to physics and engineering. They are a fundamental part of mathematics and are used to model and solve a wide variety of problems in science and technology.

11.Circles: Figures with all points at an equal distance from the center

A circle is a geometric figure that consists of a set of points in a plane that are at a constant distance from the center of the circle. This constant distance is called the radius of the circle and is represented by "r". The set of all points at a distance "r" from the center forms the edge or circumference of the circle.

Some key properties of circles include:

Center: The center of the circle is a fixed point in the plane from which all distances to the circumference are equal. The center is generally represented by the letter "O" in geometric notation.

Radius: The radius of the circle is the distance from the center of the circle to any point on the circumference. All points on the circumference are the same distance "r" from the center.

Diameter: The diameter of a circle is twice the radius. Mathematically, it is expressed as "2r". The diameter is the longest distance that can be measured within a circle and always passes through the center.

Circumference: The circumference of a circle is the length of the line that forms the edge of the circle. The formula for calculating the circumference is $C=2\pi r$, where "C" represents the circumference and "π" is the mathematical constant pi, approximately equal to 3.14159.

Area: The area of a circle is calculated using the formula $A=\pi r^2$, where "A" represents the area of the circle. This formula relates the radius of the circle to its area.

Circles are fundamental in geometry and have applications in many areas, from Euclidean geometry and trigonometry to physics, engineering, and analytical geometry. They are especially useful in representing shapes and objects in geometry and in solving geometric problems.

Circles are a fundamental geometric figure with a wide range of applications across various disciplines. Here is a more detailed description of how circles are used in different fields:

Geometry: In Euclidean geometry, circles are used to define concepts such as circumference, area, and arc length. They are also essential in constructing geometric figures and solving problems related to the relative position of points and lines.

In Euclidean geometry, circles play a central role and are used to define and explore a series of geometric concepts and properties.

Circumference: The circumference of a circle is the line that forms its outer edge. In Euclidean geometry, properties of the circumference are studied, such as its length and its relationship with the radius of the circle. The length of the circumference is calculated using the formula $C=2\pi r$, where "C" is the circumference and "r" is the radius of the circle.

Area of the Circle: The formula for calculating the area of a circle in Euclidean geometry is $A=\pi r^2$, where "A" represents the area and "r" is the radius. This formula is fundamental for determining the space enclosed by the circumference of a circle.

Geometric Construction: Circles are essential in the construction of geometric figures. They can be used to draw tangents, create right angles, divide line segments, and trace parallels, among other geometric constructions.

Relative Position of Points and Lines: Circles are used to define concepts such as the interior, exterior, and boundary of a circle, which is fundamental in classifying and describing regions and locations in Euclidean geometry.

Intersection of Circles: The intersection of two or more circles is used in solving geometric problems, such as determining common points or locating points of tangency.

Properties of the Circumference: Properties of the circumference, such as Thales' theorem and the theorem of inscribed angles, are key concepts in Euclidean geometry based on the relationship between angles and segments of the circumference.

Euclidean geometry, based on the postulates and axioms of Euclid, uses circles and their properties as an essential part of constructing figures, measuring areas and lengths, and solving geometric problems. Circles are one of the most studied and applied geometric figures in this branch of mathematics.

Trigonometry: Unit circles, which are circles with a radius of 1, are used to define trigonometric functions such as sine, cosine, and tangent. Angles in a unit circle are used to measure and calculate the trigonometric functions of angles in trigonometry.

A unit circle is a circle whose radius is equal to 1 unit. The points on the circumference of a unit circle are all at a distance of 1 unit from the center of the circle.

Angles in the Unit Circle: To relate trigonometry to the unit circle, a central angle is taken in the unit circle. The magnitude of this angle is measured in radians, where a complete angle (one full turn around the circle) is equal to 2π radians. An angle of 1 radian in a unit circle subtends an arc on the circumference of length equal to the radius, which is 1.

Trigonometric Functions: Trigonometric functions, such as sine, cosine, and tangent, are defined in relation to angles in the unit circle. For an angle in the unit circle, the sine of the angle is equal to the y-coordinate of the point on the unit circumference where the angle touches the circumference; the cosine is equal to the x-coordinate, and the tangent is equal to the ratio of sine to cosine.

Relationship with Right Triangles: The relationship between the unit circle and trigonometry also extends to right triangles. The values of trigonometric functions in the unit circle can be used to calculate side lengths and angles in right triangles.

Trigonometric Identities: The study of trigonometric functions in the unit circle leads to the development of trigonometric identities, which are equations that relate these functions and are essential in solving equations and trigonometric problems.

The use of the unit circle provides a solid geometric foundation for understanding trigonometric functions and their relationship with angles. It is a powerful tool for calculating and modeling periodic and oscillatory phenomena in mathematics, physics, engineering, and other disciplines.

Physics: Circles appear in many physical applications, from the motion of particles in circular paths to the description of waves and oscillatory movements. They are also essential in fields such as mechanics and fluid dynamics.

Circles are fundamental in physics and play a crucial role in the description and analysis of a wide variety of physical phenomena.

Kinematics: Circles are fundamental in kinematics, which is the branch of physics that studies the motion of objects. Circular motion, in which an object moves along a circular path, is modeled using concepts related to circles. This includes angular velocity, centripetal acceleration, and frequency.

Dynamics: In dynamics, which deals with forces and the motion of objects, circles are essential for understanding the motion of objects in curved paths.

Centrifugal force, which acts outward from the center of a circle, is fundamental to keeping objects in motion along a circular trajectory.

Wave Motion: Circles are used to describe sine waves, which are fundamental in the description of waves in physics. These waves have a circular shape and are used to model phenomena such as sound waves, electromagnetic waves, and oscillations in general.

Optics: Circles are used in optics to describe the reflection and refraction of light on curved surfaces, such as lenses and mirrors. Spherical mirrors and lenses are examples of surfaces that follow circular patterns.

Celestial Mechanics: In celestial mechanics, which deals with the motion of celestial bodies, circles are used to describe the orbits of planets, satellites, and comets around stars or other massive bodies.

Fluid Dynamics: In fluid dynamics, which deals with the movement of liquids and gases, circles are used to describe and model flow around solid bodies, such as wing profiles in airplanes.

Electrodynamics: In electrodynamics, which deals with electricity and magnetism, circles are used to describe the trajectories of charged particles in magnetic fields, such as in a cyclotron, a particle accelerator.

The concepts related to circles and circular trajectories are fundamental for understanding and predicting a wide range of physical phenomena in the natural world. From the motion of planets to the description of waves and the design of technological systems, circles and their properties are essential in physics and engineering.

Engineering: Circles are used in engineering to design and analyze mechanical components and systems, as well as in the manufacturing of parts and machinery that involve circular movements.

Circles are an essential tool in engineering and play a critical role in a wide variety of applications and fields of engineering.

Mechanical Design: In mechanical engineering, circles are used to design components and systems that involve circular movements. This includes the creation of shafts, bearings, gears, and sprockets, which are common components in machinery and mechanical devices.

Stress Analysis: Circles are used in stress and deformation analysis in mechanical components. The stresses in circular parts, such as disks and cylinders, can be analyzed to ensure they can withstand the loads and forces they will be exposed to.

Wheel and Tire Design: The circular shape of wheels and tires is crucial in the design of land vehicles, such as cars and bicycles, as well as in aerial and space vehicles. Proper design of wheels and tires is fundamental for stability and performance.

Part Manufacturing: In manufacturing, circles are used to machine and produce parts with circular shapes, such as gears and bearing components. Precision in manufacturing is essential to ensure the proper functioning of machinery and systems.

Machine Dynamics: In machine dynamics, circles are applied to the study of motion and vibrations in machinery and mechanical systems. This is important to ensure that systems operate in a stable and safe manner.

Aerospace Engineering: In aerospace engineering, circles are used in the construction of components and structures, such as aircraft wings and rocket engines. The shape and arrangement of circular components are critical in aerodynamics and propulsion.

Electronic Engineering: Circles are also used in electronic engineering in the creation of printed circuits and circuit boards, which often have circular shapes. This is fundamental in electronics, where efficient circuit board design is essential.

In summary, circles are an essential tool in mechanical, civil, aerospace, electronic engineering, and many other disciplines. They are used in the design, analysis, and manufacturing of components and systems, as well as in the creation of structures and devices that involve circular movements. Understanding and applying concepts related to circles are crucial for success in engineering.

Analytical Geometry: In analytical geometry, circles are described using mathematical equations that relate the coordinates of their points. This allows for analyzing and solving geometric problems using algebra and calculus.

Indeed, analytical geometry is a branch of mathematics that combines classical geometry with algebra and calculus and allows for the description of geometric figures, including circles, through mathematical equations. In the case of circles, the description through equations provides a precise way to represent and analyze their position and properties in a coordinate system.

Intersection Analysis: The equations of circles are used to analyze intersections and relationships between circles and other geometric figures, such as straight lines. For example, two circles may be tangent to each other or may intersect at two points.

Solving Geometric Problems: Analytical geometry allows for solving geometric problems related to circles, such as finding the intersection point between a line and a circle, determining if a point is inside or outside a circle, or finding the equation of a circumference that passes through three given points.

Geometric Transformations: Geometric transformations, such as translation, rotation, and reflection, can be described and analyzed using equations in analytical geometry. These transformations can be applied to circles to modify their position and orientation.

Study of Geometric Properties: Analytical geometry is also used to study geometric properties of circles, such as their diameter, circumference, area, and the relative position of various circles in a plane.

Analytical geometry provides a powerful tool for tackling geometric problems algebraically, facilitating the solving of equations and the manipulation of variables in geometric contexts. This is valuable in mathematics and practical applications in engineering, physics, and other fields where precise and rigorous geometric analysis is required.

Computer Graphics: Circles are used in computer graphics for representing graphics and designing objects in three-dimensional models, as they are fundamental in creating curves and smooth surfaces.

Circles and related curves are fundamental components in the field of computer graphics. The representation and manipulation of circles, as well as the creation of curves and smooth surfaces, are essential for generating computer graphics and designing objects in three-dimensional models.

Drawing Circles and Arcs: In computer graphics applications, algorithms are used to draw circles and arcs accurately on screens or surfaces. These algorithms ensure that circles are represented smoothly and visually accurately.

Modeling Objects and Surfaces: In creating three-dimensional models, circles and related curves are used to design objects and smooth surfaces. For example, they can be used to create profiles of three-dimensional objects or to define contours in character and object modeling.

Animation: Circles and curves are used in computer animation to define motion trajectories. Bézier curves and splines are particularly popular for describing smooth movements in animations.

Collision Detection: Circles are used in collision detection algorithms in video games and simulations to determine if two moving objects intersect. This information is crucial for interaction and real-time physics.

Rendering Curved Surfaces: In rendering, techniques for rendering curved surfaces, such as NURBS (Non-Uniform Rational B-Splines) surfaces, are used, which are useful for representing objects with smooth and complex surfaces.

Graphic Design and CAD: In graphic design and computer-aided design (CAD), circles and curves are employed to create illustrations, logos, blueprints, and detailed designs. Precision in the representation of curves is fundamental in these contexts.

Image Editing: In image editing applications, tools are used that allow for editing and manipulating curves and selections. These tools can be useful for tracing circular shapes or applying effects to images.

In summary, circles and curves are essential components in computer graphics and are applied in a variety of contexts, from drawing and modeling objects to animation and rendering. The ability to represent smooth and precise curves is crucial for achieving realistic and visually appealing results in computer graphics applications.

Topography: Circles are used in topography to describe and measure contour lines on topographic maps and in determining elevations and slopes on terrains.

In topography, the representation and analysis of circles and related curves are fundamental for the detailed description of the Earth's surface. This is achieved through the use of contour lines on topographic maps.

Contour Lines: Contour lines are lines on a topographic map that connect points with the same elevation above sea level. These lines resemble concentric circles and are used to represent the three-dimensional shape of the terrain on a two-dimensional plane. Contour lines allow surveyors and cartographers to effectively show the elevation and slope of the terrain.

Measuring Elevations: The elevation of a point on the terrain can be determined by following a contour line to the intersection point with a known elevation line (often called a "reference contour"). This method of elevation measurement is essential in topography and cartography.

Calculating Slopes: Slopes on the terrain are calculated from the horizontal distance between contour lines and the elevation difference between them. This is useful for understanding the inclination of the terrain and its importance in planning projects, such as roads, railways, and dams.

Infrastructure Design: Engineers and planners use topographic maps with contour lines to design infrastructure projects. Accurate representation of the terrain is essential to ensure that constructions fit the landscape and are built safely and effectively.

Navigation and Orientation: Topographic maps are essential for land navigation and outdoor orientation. Contour lines provide detailed information about the topography and allow hikers, geologists, and outdoor activity professionals to better understand the geography of an area.

Geological and Environmental Studies: Circles and contour lines are valuable tools in geological and environmental studies. They allow scientists and researchers to characterize and analyze the terrain and its impact on geology and the environment.

Natural Resource Management: Topographic maps and contour lines are used in natural resource management, such as land use planning, forest management, and assessing flood-prone areas.

In summary, in topography, circles are used through contour lines to provide a detailed representation of the Earth's topography. This is fundamental for a variety of applications, from planning construction projects to outdoor navigation and natural resource management.

Navigation: In navigation, circles are used to calculate distances on maps and to determine location and direction in maritime and aerial navigation.

The use of circles in navigation is a common and fundamental practice for determining location, direction, and calculating distances in maritime, aerial, and land navigation.

Circles of Longitude and Latitude: The grid of lines of longitude and latitude that covers the Earth's surface forms a kind of grid with circles. Latitude circles are parallel to the equator, and longitude circles are semicircles that go from the North Pole to the South Pole. The intersection of these circles provides geographic coordinates used to identify locations on Earth.

Determining Position: Navigators use circles of longitude and latitude to determine their position at sea or in the air. This is done by observing the position of the sun, stars, or navigation satellites such as GPS. By comparing the observed coordinates with those recorded on maps and navigation systems, the exact location can be determined.

Calculating Distances: Circles are also used to calculate distances on maps, especially in maritime navigation. The distance between two points on the Earth's surface can be determined using circles of longitude and latitude and applying trigonometric formulas.

Celestial Navigation: In maritime and aerial navigation, position can be determined through observations of celestial bodies, such as the Sun, Moon, planets, and stars. These observations are made in relation to the observer's position on Earth, using circles of longitude and latitude.

Compass and Bearings: Navigators use compasses to determine directions and bearings. The circles of a compass are used to measure direction in degrees relative to cardinal points. This information is crucial for maintaining the correct course during a journey.

Route Planning: Circles and position information are used in route planning in navigation. Navigators plot routes on maps, taking into account the location of obstacles, weather conditions, and other factors that may affect navigation.

Electronic Navigation: In modern navigation, electronic navigation systems such as GPS are used, which are based on satellites to provide precise coordinates and determine position and direction.

Circles are an essential part of navigation and provide the basis for determining location, direction, and distance in land, maritime, and aerial environments. The ability of navigators to understand and apply concepts related to circles is fundamental for safe and effective navigation.

Circles are a versatile geometric figure that applies to a wide variety of disciplines and fields, from pure mathematics to physics, engineering, and computer science. Their simplicity and mathematical properties make them fundamental tools in solving geometric problems and in representing phenomena that involve circular movements and relationships.

12.Area: The amount of space inside a figure, like a square or a circle

The area is a measurement that quantifies the amount of two-dimensional space occupied by a figure on a plane. The concept of area applies to a wide variety of geometric figures, such as squares, circles, triangles, rectangles, polygons, and other shapes. The unit of measurement for area is usually related to the unit of measurement for length, but squared (for example, square meters, square centimeters, or acres).

Area of a Square: The area of a square is calculated by multiplying the length of one of its sides by itself. In mathematical notation, if l is the length of a side of the square, then the area A is expressed as $A = l^2$.

Area of a Rectangle: The area of a rectangle is c calculated by multiplying the length of one of its sides (base) by the length of the other side (height). Mathematically, if b is the base and h is the height, then the area A is expressed as $A = b \cdot h$.

Area of a Triangle: The area of a triangle is calculated by multiplying the length of the base by the height and dividing the result by 2. Mathematically, if b is the base and h is the height, then the area A is expressed as $A = \frac{1}{2} \cdot b \cdot h$.

Area of a Circle:

Have you ever looked at a whole pizza and wondered how much "total pizza" there is? That amount depends on the area of the circle, because a pizza, like many round objects, is shaped like a circle.

What is area?

Area is simply the space a shape takes up on a flat surface. In other words: if you could fill a circle with tiny squares, the area tells you how many fit inside.

How do you calculate the area of a circle?

Easy! You only need one thing: the radius of the circle.

The radius is the distance from the center of the circle to the edge.

Once you have the radius, use this magic formula:

$$\text{Area} = \pi \times r^2$$

Where:

π (pi) is a special number that's always approximately 3.1416 (don't worry, you don't need to memorize all its infinite decimals).

r is the radius of the circle.

r2r^2r2 means "radius squared", or radius times radius.

Simple example:

Imagine a circle with a radius of 3 centimeters.

Area=π×32=π×9≈3.1416×9=28.27 cm2\text{Area} = \pi \times 3^2 = \pi \times 9 \approx 3.1416 \times 9 = 28.27 \text{ cm}^2Area=π×32=π×9≈3.1416×9=28.27 cm2

Done! That means the circle takes up about 28.27 square centimeters.

Tip so you don't forget:

Think of "pirates" to remember the formula:

"Pi times r squared" → sounds like "pirate squared

Area of Irregular Figures: For figures with irregular shapes, such as polygons, they can be divided into simpler parts (triangles, rectangles, etc.), and then the areas of these parts can be summed to find the total area of the figure.

Practical Applications: The calculation of area is applied in a variety of contexts, from construction and architecture, where the area of surfaces and land is calculated, to computational geometry, where the area of figures in digital environments is determined. It is also fundamental in fields such as physics, where the area of surfaces and regions is measured in calculations of density and flow.

The calculation of area has a wide range of practical applications in various fields. Below are some areas where the calculation of area is fundamental:

Construction and Architecture: In construction and architecture, the area of surfaces, such as walls, roofs, and floors, is calculated to determine the amount of materials needed. The area is also used in the planning and design of buildings and structures. The calculation of area plays a fundamental role in various stages of the process, from planning and design to estimating materials and costs.

Design and Planning: In the design phase of a building or structure, area calculation is essential for determining the distribution of spaces, the location of rooms, the arrangement of furniture, and optimizing the overall design. Architects use area to define the functionality of a space and create detailed plans.

Material Estimation: Once a project has been designed, the area of the involved surfaces, such as walls, roofs, and floors, needs to be calculated.

This is fundamental to determine the amount of materials needed, such as bricks, blocks, wood, drywall, tiles, or flooring. Accurate area calculation ensures that materials are acquired in the correct quantity, avoiding waste and unnecessary costs.

Costs and Budgeting: Area is also used in estimating costs and construction budgets. The cost of materials and labor is directly related to the area to be covered. Contractors and construction companies use area calculations to determine the budget and scope of a project.

Paint Surface Calculation: In painting projects, whether indoors or outdoors, the area of the surfaces to be painted is calculated. This allows for determining the amount of paint needed for adequate coverage and a uniform finish.

Cladding Planning: In the selection of cladding materials, such as tiles, wallpaper, or flooring, the area of the surfaces influences the amount of material that needs to be purchased. Area calculation is important to ensure that enough cladding is acquired and to avoid shortages or excess material.

Compliance with Codes and Regulations: Building codes and regulations often specify requirements related to area, such as height and minimum space allowed in buildings. Architects and builders must comply with these standards to ensure safety and regulatory compliance.

Space Optimization: Area calculation is also used to optimize the use of spaces in residential and commercial projects. The aim is to maximize the available space and ensure efficient distribution.

Renovation and Expansion: In renovation and expansion projects, area calculation is essential to determine the necessary modifications and adjustments to existing structures. It is also used to estimate costs and project planning.

In summary, area calculation plays a crucial role in construction and architecture, as it directly affects design, material estimation, costs, and project planning. Accuracy in area calculation is essential for the success of a construction project and for ensuring that resources are used efficiently.

Topography and Cartography: In topography, the area of land and regions on topographic maps is calculated. Contour lines are used to measure land areas and calculate distances on maps.

In topography and cartography, area calculation plays a crucial role in the accurate representation of the Earth's surface and the creation of topographic maps. Contour lines and other techniques are used to measure land areas and calculate distances on maps.

Contour Lines: Contour lines are lines that connect points on the terrain with the same elevation above sea level. These lines are represented on topographic maps and are essential for visualizing the topography of a region. Each contour line is labeled with its elevation in meters or feet. The distance between contour lines indicates the steepness of the terrain: the closer the lines are, the steeper the terrain.

Area Calculation: To calculate the area of a region on a topographic map, contour lines are used. Once the elevation of the contours delimiting the area of interest is determined, area calculation can be utilized to measure the extent of the land in question. This is important in topography and cartography to quantify land areas, watersheds, parcels, geographical zones, and more.

Distance Calculation: In addition to area, contour lines are also used to calculate distances on a topographic map. Since each contour line is related to a specific elevation, it is possible to measure vertical and horizontal distances accurately. This is crucial for planning and navigation in mountainous and rugged terrains.

Geodetic Studies: Topography relies on precise measurements of the Earth. Surveyors use geodetic coordinate systems and advanced techniques to perform exact measurements and determine areas with high precision. This is essential in engineering projects, urban planning, infrastructure construction, and more.

Elevation Mapping: Calculated areas and topographic measurements are used to create elevation maps, which show the detailed topography of an area. These maps are valuable for planning hiking routes, conducting environmental impact studies, analyzing flood-prone areas, and more.

Project Planning: In both topography and cartography, area calculation and distance measurement are fundamental for project planning. Engineers, urban planners, and infrastructure designers use topographic maps to assess the feasibility of projects and understand the terrain in which they will work.

Environmental Studies: In environmental cartography and topography, area and distance measurements are used to perform environmental impact assessments, land use studies, and monitor changes in the landscape.

Area calculation and topographic measurements are essential in topography and cartography to accurately understand and represent the Earth's surface. These techniques are valuable for a wide range of applications, from construction project planning to environmental management and navigation in diverse terrains.

Computational Geometry: In computational geometry, the area of geometric figures in digital environments is determined. This is essential in computer graphics, image processing, and computer-aided design (CAD).

In computational geometry, area calculation plays a crucial role in the representation and analysis of geometric figures in digital environments. This is fundamental in fields such as computer graphics, image processing, and computer-aided design (CAD).

Computer Graphics: Area calculation is essential in the creation and rendering of computer graphics. It is used to determine the area of surfaces, such as polygons and three-dimensional meshes, that form 3D models. This is crucial for generating realistic images in video games, movies, and 3D modeling applications.

Image Processing: In image processing, area calculation is used to measure regions of interest in a digital image. This is useful in applications such as object segmentation, edge detection, and measuring areas of specific regions in an image.

Computer-Aided Design (CAD): In computer-aided design, engineers and designers use CAD software to create models of products and structures. Area calculation is used to determine areas of cross-sections, surfaces, and components of designs.

Computational Geometry in Robotics: In robotics, computational geometry is applied to determine the areas of robot workspaces and plan routes. Area calculation is used to assess accessibility and collision in three-dimensional space.

Geospatial Data Visualization: In geographic information systems (GIS) and geospatial data visualization, area calculation is used to measure areas of land parcels, geographical regions, and political boundaries on digital maps.

Terrain Modeling: In terrain modeling applications, such as creating digital elevation models (DEMs), area calculation techniques are used to measure the extent of geographical features, such as watersheds, mountains, and valleys.

Scientific Simulations: In scientific simulations, area calculation is used to analyze and measure areas of regions of interest in scientific data. This is applicable in fields such as meteorology, oceanography, and particle physics.

Data Analysis and Statistics: In data analysis and statistics, area calculation is useful for measuring areas under curves in graphs, providing insights into the distribution and relationship between variables.

Object Recognition and Computer Vision: In the field of computer vision, area calculation is used in applications such as object recognition and detecting areas of interest in images and videos.

Area calculation in computational geometry allows for measuring and analyzing geometric figures in digital environments, which is essential in numerous applications, ranging from graphics creation and 3D modeling to image processing and simulation in various scientific and technological fields.

Physics: In physics, area is measured in contexts such as density calculations, field flow, and force distribution. Area is an important variable in physical equations involving surfaces and regions.

In the field of physics, area calculation is fundamental in a variety of contexts and applications to understand and quantify properties of surfaces and regions.

Density: Area calculation is used to determine the surface density of an object or substance. Surface density is mass per unit area and is applied, for example, in geophysics to study the density of rocks and minerals in the Earth.

Field Flow: In physics, especially in electrostatics and magnetism, Gauss's theorem is based on calculating the flow of a field through a closed surface. The area calculation of the surface is essential to determine the flow of electric or magnetic fields through that surface. This theorem is fundamental to understanding the behavior of electric charges and magnetic fields.

Force Distribution: In mechanics and dynamics, area calculation is used to analyze the distribution of forces over a surface. For example, in structural engineering, stresses in a structure can be calculated by measuring the force distribution across a surface.

Optics: In optics, area calculation is relevant for determining the amount of light that strikes a surface and the intensity of luminous radiation over a specific area. This is important in lighting, lens optics, and the description of optical systems.

Thermodynamics: In thermodynamics, area calculation is also relevant, especially when working with property diagrams, such as the phase diagram of a fluid. The area under a curve in these diagrams provides information about thermodynamic properties, such as enthalpy or entropy.

Fluid Mechanics: In fluid mechanics, area calculation is used to analyze flow speed through a surface. This is crucial in applications such as aerodynamics, hydrodynamics, and the design of ducts and pipes.

Solar Radiation: In solar radiation studies, such as in photovoltaic solar energy, area calculation is performed to determine the amount of energy that can be captured by solar panels based on their surface area.

Electric Charge Distribution: Area calculation is applied in problems of electric charge distribution on plates and capacitors, where the area of the plates is measured to determine electrical capacitance.

Area calculation in physics allows for quantifying properties that are essential for understanding and predicting the behavior of physical systems across various disciplines, from thermodynamics to optics and electronics. Area is used to analyze the relationship between surface properties and physical properties in scientific and technological applications.

Civil Engineering: In civil engineering, area calculation is essential for the design of infrastructure, such as roads, bridges, and dams. It is used in land planning and environmental impact analysis.

Civil engineering is a field in which area calculation plays a vital role in the planning, design, and construction of a wide range of infrastructure projects.

Road and Street Design: In the planning and design of roads and streets, area calculation is used to determine the amount of land that needs to be excavated or filled to level the terrain. Area calculation is also essential for designing the cross-sections of roadways and determining the width of roads and sidewalks.

Bridge Design: In bridge engineering, area calculation is applied to design the piers and abutments of bridges, as well as to determine the amount of materials needed for their construction. It is also used in analyzing loads and stresses on bridge structures.

Dams and Hydraulic Works: In the construction of dams and hydraulic works, area calculation is fundamental for determining water storage capacity and assessing pressure distribution in structures. It is also used in the design of channels and dikes.

Urban Development Planning: In urban development planning and building construction projects, area calculation is applied to determine the total area of available land and to calculate the areas of parcels and lots. This is important in land subdivision and urban zoning.

Natural Resource Management: Area calculation is used in the management of natural resources, such as evaluating agricultural land, forest planning, and identifying conservation areas. It allows for quantifying areas of specific use and evaluating resource distribution.

Environmental Impact Analysis: In civil engineering projects, an environmental impact assessment is carried out to evaluate the effect of works on the natural environment. Area calculation is fundamental to measure the impact on green areas, water bodies, and ecologically sensitive zones.

Topography and Leveling: Topography and leveling are techniques that use area measurements to determine elevations and terrain profiles. This is essential in civil engineering to ensure the safety and stability of structures and roads.

Volume Determination: Area calculation is a preliminary step for calculating volumes of earth moved in construction projects, such as excavations or embankments. This information is important for cost estimation and project scheduling.

Load Evaluation: Area calculation is also used in load evaluation on structures and foundations. Determining the contact area between a structure and the ground is essential for predicting stresses and structural response.

In summary, area calculation is an essential tool in civil engineering for the design, planning, and construction of infrastructure projects. It allows for evaluating the dimensions of land, structures, and construction elements, as well as quantifying resources and assessing environmental impact, which is crucial for the efficient management of projects and the creation of safe and sustainable environments.

Agriculture and Forestry: In agriculture and forestry, the area of fields, forests, and plots is measured to manage resources, calculate yields, and plan planting.

In agriculture and forestry, area calculation plays an essential role in managing natural resources and planning activities. Here are some key applications of area calculation in these fields:

Agriculture:

Crop Management: Farmers use area calculation to determine the size of crop fields. This is important for planning the amount of seeds, fertilizers, and other inputs needed. It also allows for estimating expected production.

Irrigation and Fertilization: Area calculation is essential for determining the amount of irrigation water and fertilizers needed to uniformly cover a crop field. Resource efficiency is key to sustainable agriculture.

Crop Diversification: Farmers can use area calculation to plan the diversification of crops on their land. This contributes to crop rotation, which can improve soil health and prevent diseases.

Pest and Disease Control: Area calculation is also useful in identifying areas affected by pests or diseases. It allows for delineating treatment and monitoring zones.

Harvest Planning: Area calculation is essential for planning the harvest and determining the right time for crop collection.

Forestry (Forest Management):

Forest Inventory: Forestry professionals use area calculation to measure the extent of forests and forest plots. This is essential for inventorying timber resources and planning their sustainable exploitation.

Felling Planning: Area calculation is crucial in planning tree felling. It allows for estimating the amount of available wood and ensuring compliance with sustainable forest management regulations.

Reforestation: For reforestation or tree planting projects, area calculation is important for determining the number of trees needed and distributing them evenly across a plot.

Biodiversity Studies: Area calculation is also used in biodiversity studies to define research and conservation areas in forests and natural ecosystems.

Environmental Impact Assessment: In forestry, area calculation is relevant in assessing the environmental impact of activities such as building forest roads or resource exploitation projects.

Yield Determination: In both fields, area calculation is fundamental for determining yields, whether agricultural production per hectare or the amount of wood per acre.

Area calculation is an essential tool in agriculture and forestry for effectively planning and managing natural resources, optimizing production, and ensuring sustainable practices that respect the environment and biodiversity.

Biology: In biology, area is calculated for geographical regions to study habitats and ecosystems. Area is an important variable in ecology and natural resource management.

In biology, area calculation is a valuable tool for understanding and studying ecosystems, habitats, and species distribution.

Ecology and Habitat Study: Ecologists use area calculation to define and measure habitats and ecosystems. This allows for quantifying the extent of a study area, which is fundamental for analyzing species distribution, biodiversity, and interactions between organisms and their environment.

Biodiversity Conservation: In biodiversity conservation, area calculation is used to identify and delineate protected areas, such as national parks and nature reserves. It is also essential for assessing the effectiveness of these areas in preserving endangered species.

Landscape Study: Scientists use area calculation to analyze the structure of landscapes and ecosystem fragmentation. This is relevant for understanding how changes in land use and urbanization affect ecosystems and habitat connectivity.

Species Distribution Modeling: In species distribution studies, the area of a species' geographical distribution is calculated. This helps identify areas where a species is more likely to be found and provides valuable information for conservation and resource management.

Microhabitat Studies: Area calculation is used to measure microhabitats or smaller areas within an ecosystem, such as bird nests, animal burrows, and specific patches of vegetation.

Environmental Impact Assessment: In environmental impact assessments of development projects, area calculation is used to evaluate how human activities may affect local ecosystems and biodiversity.

Natural Resource Management: In natural resource management, area calculation is performed to assess the sustainability of resource exploitation, such as fishing, hunting, and logging. This is important to ensure the long-term conservation of natural resources.

Ecological Corridor Analysis: Ecological corridors are areas that connect fragmented habitats, allowing species to move and reproduce. Area calculation is relevant for determining the extent and effectiveness of these corridors in biodiversity conservation.

Climate Change Studies: In the context of climate change, area calculation is applied to assess the effects of global warming on species distribution and migration patterns.

Area calculation in biology is essential for delineating study areas, assessing species distribution, conserving biodiversity, and understanding how changes in the environment can impact ecosystems and wildlife. This tool is fundamental for natural resource management and environmental conservation.

Economics: In economics, area is calculated under curves in graphs to represent and analyze data. This is used in economic and financial analysis.

In the field of economics, calculating the area under curves in graphs plays a crucial role in representing and analyzing data, especially in economic and financial analysis. Below are some key applications of area calculation in economics:

Data Representation: Area calculation is used to represent economic data in graphs, such as bar graphs, line graphs, and stacked areas. This provides a visual representation of economic information, facilitating the understanding and communication of trends and patterns.

Time Series Analysis: In the analysis of economic time series, area calculation under the curves of lines or areas is employed to evaluate trends over time. This is important for understanding the evolution of economic variables, such as GDP growth, unemployment, and inflation.

Financial Performance Evaluation: In financial analysis, area calculation is applied to graphs of asset performance, investment portfolios, and mutual funds. It helps investors and analysts evaluate historical performance and compare different investment options.

Market and Demand Measurement: Area calculation under supply and demand curves is used in economics to measure the quantity of goods and services traded in a market and to analyze demand elasticity. This is essential in microeconomics and business decision-making.

Elasticity Studies: Demand elasticity and supply elasticity are fundamental concepts in economics. Area calculation is used to measure the percentage change in quantity demanded or supplied in response to changes in prices or income.

Production Economics: Area calculation is applied in production theory to analyze the relationship between production factors (labor and capital) and total output. It allows for determining productivity and efficiency in a firm.

Valuation of Options and Financial Derivatives: In finance, area calculation under curves is relevant in valuing options and financial derivatives. Valuation formulas, such as the Black-Scholes formula, involve calculating areas under probability distribution curves.

Health Economics: In health economics, area calculation is used to evaluate the cost and effectiveness of medical treatments and health programs. It is also relevant in cost-effectiveness studies and decision analysis in the healthcare sector.

Cost-Benefit Analysis: In project and public policy analysis, area calculation is used to evaluate the costs and benefits of different alternatives. This is fundamental in governmental decision-making and investment project evaluation.

Area calculation in economics provides a powerful tool for analyzing data, evaluating economic trends, making financial decisions, and conducting economic research. It allows for quantifying changes and relationships in a variety of economic contexts, which is essential for decision-making and policy formulation.

Graphic Design: In graphic design and visual arts, area calculation is used in creating illustrations and visual compositions. Area is considered in the arrangement of elements in a composition.

In graphic design and the visual arts, area calculation plays a fundamental role in creating visually appealing and effective compositions. Here are some key applications of area calculation in graphic design and visual arts:

Composition and Page Design: Graphic designers use area calculation to distribute visual elements on a page, whether in print or digital media. This includes arranging text, images, graphics, and other elements to achieve a balanced and visually attractive presentation.

Logo and Brand Design: In creating logos and brands, area calculation is essential to ensure that the design is coherent and proportionate. It ensures that logo elements occupy space evenly and fit the desired proportions.

Advertising Design: In advertisements and promotional materials, area calculation is applied to design elements such as banners, posters, and print or digital ads. This contributes to the effectiveness of advertising by grabbing attention and conveying messages clearly.

Packaging and Label Design: In designing product packaging and labels, area calculation is important to ensure that information, graphic design, and visual elements fit the available space on the package.

Illustration and Digital Art: Illustrators and digital artists use area calculation when creating illustrations and works of art. This applies to distributing elements on a digital canvas and controlling the composition of the artwork.

Web and User Interface Design: In web and user interface design, area calculation is essential for organizing elements such as buttons, images, headers, and text on a webpage or application. It contributes to usability and user experience.

Magazine and Book Design: In magazine and book layout, area calculation is used to determine the location and size of elements such as text columns, images, margins, and headers. This ensures a visually attractive and readable presentation.

Exhibition and Museum Design: In exhibition and museum design, area calculation is applied when planning the arrangement of artworks, exhibits, and interactive elements in an exhibition space.

Scenography and Theater Design: In scenography and theater design, area calculation is relevant for arranging stages, sets, and props in a performance space.

Area calculation in graphic design and visual arts is an essential tool for creating effective, balanced, and aesthetically pleasing visual compositions. It allows designers and artists to carefully control the arrangement of elements and the distribution of space, which is crucial for achieving the desired visual impact and effectively communicating messages.

Astronomy: In astronomy, area calculation is used to measure areas in the sky, such as the surface of planets or the magnitude of areas of constellations.

In astronomy, area calculation plays an important role in measuring areas in the sky and studying celestial objects. Here are some key applications of area calculation in astronomy:

Surface of Planets and Moons: Astronomers and planetary scientists use area calculation to measure the surface of planets, moons, and other celestial bodies. This is essential for determining their surface area, calculating the extent of geological features, and studying the topography of these bodies.

Magnitude of Areas in the Sky: In celestial cartography and constellation identification, area calculation is employed to measure the magnitude of areas in the sky. This is important for determining the apparent extent of constellations and regions of celestial space.

Solid Angle Calculation: In astronomy, astronomers use area calculation to measure solid angles in celestial space. This is relevant in determining viewing angles, telescope coverage areas, and observations of objects in the sky.

Lunar Surface Mapping: For studies of the Moon, such as lunar cartography and space exploration, area calculation is used to map the lunar surface and determine the extent of craters, lunar seas, and other features.

Modeling Planetary Surfaces: Area calculation is important in modeling planetary and lunar surfaces in 3D. This allows for creating accurate representations of the topography of planets and moons for research and space exploration purposes.

Size and Distance Determination: By combining area measurements with information about the distance to celestial objects, astronomers can calculate relative sizes and distances in space, which is fundamental for positional astronomy and parallax determination.

Study of Occultations and Eclipses: Area calculation is used to study astronomical events such as occultations and eclipses. It helps predict when and where these phenomena will occur and understand their duration and extent.

Study of Deep Sky Areas: In astrophotography and observation of deep sky objects, area calculation is relevant for measuring the apparent magnitude of starry sky areas that contain galaxies, nebulae, and star clusters.

Area calculation in astronomy is an essential tool for the precise measurement of celestial surfaces, analysis of observational data, and study of astronomical objects and phenomena. It allows astronomers to quantify and compare features in space and contributes to our understanding of the universe.

In conclusion, area calculation is a versatile mathematical tool that is applied in a wide variety of fields and disciplines to solve problems, make decisions, and perform accurate measurements. Its importance lies in its ability to quantify the extent of surfaces and regions, which is essential in everyday life and in the research and application of knowledge across numerous fields. The concept of area is fundamental in geometry and many other disciplines, as it provides a way to quantify the extent of a figure in two dimensions and is crucial for solving mathematical problems and practical applications in everyday life and various areas of knowledge.

13. Perimeter: The distance around a figure

The perimeter is the distance that surrounds or encloses a geometric figure, such as a polygon, a circle, or any two-dimensional shape. The perimeter is calculated by adding the lengths of all the sides of the figure. Depending on the specific shape of the figure, the calculation of the perimeter may vary.

Perimeter of a Polygon: To calculate the perimeter of a polygon, you sum the lengths of all its sides. For example, in a triangle, you would add the lengths of the three sides. In a square, you would add the four sides.

To calculate the perimeter of a polygon, you simply add the lengths of all its sides. Here is an example to illustrate how to calculate the perimeter of a triangle and a square:

Example 1: Calculation of the Perimeter of a Triangle

Let's say you have a triangle with sides of lengths "a", "b", and "c". The perimeter (P) of the triangle is calculated as:

$P = a + b + c$

You simply add the lengths of the three sides of the triangle to obtain the perimeter.

Example 2: Calculation of the Perimeter of a Square

A square has all its sides of equal length. Let's say the length of one side of the square is "s". The perimeter (P) of the square is calculated as:

$P = 4s$

In this case, you multiply the length of one side by 4, since the square has four equal sides.

This same approach applies to any polygon. To calculate its perimeter, you only need to know the lengths of its sides and sum them. The perimeter is a fundamental measure in geometry and is used in a variety of applications, from construction to design and the calculation of fences and edges in various projects.

Perimeter of a Circle: In the case of a circle, the perimeter is known as the circumference. The circumference is calculated using the formula $2\pi r$, where "r" is the radius of the circle. This means that the circumference is approximately 6.28 times the length of the radius.

To calculate the circumference of a circle, the formula you mentioned is used: $2\pi r$, where "r" represents the radius of the circle. The constant π (pi) is an irrational number that is commonly approximated as 3.14159, although it can be used with greater precision in more detailed calculations.

The complete formula for calculating the circumference (C) of a circle is:

$C = 2\pi r$

Where:

"C" is the circumference of the circle.

"π" is the constant pi (approximately 3.14159).

"r" is the radius of the circle, which is the distance from the center of the circle to any point on its edge.

This formula is fundamental in geometry and is used in a variety of applications in mathematics and practical fields such as engineering, physics, cartography, and computational geometry to calculate arc lengths, perimeters of circular figures, and other measurements related to circles.

Perimeter of an Irregular Shape: For irregular shapes, you can calculate the perimeter by dividing the figure into smaller segments and summing the lengths of those segments. This is common in computational geometry, where complex shapes are divided into simpler segments to facilitate the calculation of the perimeter.

The calculation of the perimeter of irregular shapes generally involves dividing the figure into simpler segments and then summing the lengths of those segments to obtain the total perimeter. This approach is common in computational geometry and in situations where the figures do not have a regular or known geometric shape.

Let's say you have an irregular figure, like the one shown in an image or in a textual description. To calculate its perimeter, you can follow these steps:Divide the Figure: Observe the figure and divide its outline into smaller segments. You can use straight lines to connect points on the figure's edge.

Measure the Segments: Use a ruler or measuring instrument to determine the length of each segment. Record these lengths.

Sum the Lengths: Add all the lengths of the segments to obtain the total perimeter of the figure.

This approach is especially useful when working with complex figures or shapes that cannot be easily described using simple mathematical formulas. In practice, dividing the figure into smaller segments and measuring their lengths can be done manually or with the help of computational geometry software to facilitate the calculation of the perimeter.

The calculation of the perimeter is important in geometry, architecture, construction, and many other disciplines where measuring or determining the distance around a figure is required. It is also essential for estimating materials, such as fences or edges, in construction and design projects.

The calculation of the perimeter is a fundamental measure in geometry and has significant applications in a variety of disciplines and fields.

Geometry: The calculation of the perimeter is essential for the description and comparison of geometric figures. In geometry, it is used to determine the length of the outline of a figure, which is fundamental for understanding its properties and relationships with other figures.

From Figures: The calculation of the perimeter helps identify and distinguish between different types of geometric figures. For example, the length of the outline of a square is different from that of a triangle, which allows for the classification and recognition of these figures.

Comparison of Sizes: The calculation of the perimeter is used to compare the relative size of figures. You can determine which of two figures has a greater or lesser perimeter, which is fundamental for solving geometric problems and understanding concepts of area and proportion.

Calculation of Distances: In real-world situations, such as measuring the walls of a room or the perimeter of a piece of land, the calculation of the perimeter is essential for determining distances and lengths.

Properties of Composite Figures: When it comes to composite figures or shapes that consist of several parts (such as an irregular polygon), the calculation of the perimeter is used to find the total length of the shape's outline.

Study of Proportions: The calculation of the perimeter is relevant for understanding the relationships of proportion between figures. For example, if you have two rectangles with a specific relationship between their lengths and widths, their perimeters will also have a proportional relationship.

Solving Geometry Problems: In geometric problems, especially those that involve figures and measurements, the calculation of the perimeter is a common operation to determine precise answers.

Design and Construction: In architecture and design, the calculation of the perimeter is applied in the planning and construction of structures and buildings. Architects and builders use perimeter measurements to determine the amount of material required and the design of the structures.

The calculation of the perimeter is an essential skill in geometry that applies to various situations for describing figures, solving problems, and understanding geometric properties. It helps mathematicians, students, and professionals analyze and relate different geometric figures accurately and effectively.

Architecture and Construction: In architecture and construction, the calculation of the perimeter is used to measure the length of walls, fences, buildings, and other structures. It is essential for determining the amount of materials needed, such as bricks, concrete, or fencing, and for estimating construction costs.

The calculation of the perimeter plays a critical role in architecture and construction, and its importance spans numerous aspects in these fields.

Design and Planning: Architects use the calculation of the perimeter to determine the total length of the walls of a building or structure. This is essential in the design and planning phase as it allows for proper sizing of spaces and determining the amount of materials needed.

Material Estimation: By calculating the perimeter of a structure, construction professionals can estimate the amount of required materials. This includes bricks, concrete blocks, wood, steel, or other building materials. An accurate estimation of the perimeter is crucial to avoid shortages or excess materials.

Construction Budgeting: The calculation of the perimeter contributes to the construction budget. By estimating the amount of materials needed and knowing the cost of each material per unit, contractors can determine the total costs of construction materials.

Design of Fences and Boundaries: In projects involving fences, retaining walls, and property boundaries, the calculation of the perimeter is fundamental. It helps define the length of fences and walls, which is relevant for property delineation and securing private areas.

Determination of Areas to Paint or Cover: In the application of paint or covering on structures, the calculation of the perimeter is used to determine the area that needs to be covered. This is essential for calculating the amount of paint or covering needed.

Blueprint Verification: During construction, architectural and construction blueprints are used to guide the process. The calculation of the perimeter is a way to verify that actual dimensions match the specifications of the blueprint.

Planning of Interior Spaces: The calculation of the perimeter of the interior walls of a building is relevant for the distribution of interior spaces and the arrangement of elements such as furniture and partitions.

Safety and Regulations: Measuring the perimeter is also important to ensure that structures comply with building codes, including safety and access requirements.

The calculation of the perimeter is a crucial tool in architecture and construction. It allows construction professionals to plan, budget, and execute projects efficiently, ensuring that structures are designed and built accurately while complying with applicable regulations and standards.

Garden Design and Landscaping: In landscaping projects, the calculation of the perimeter is used to measure the length of gardens, paths, and landscape features. It aids in the planning and design of outdoor spaces.

The calculation of the perimeter plays an important role in garden design and landscaping, as it contributes to the planning and creation of attractive and functional outdoor spaces.

Design of Gardens and Green Areas: By measuring the perimeter of an area designated for gardens, parks, or green spaces, landscape designers can determine the amount of plants, shrubs, and grass required. This is crucial for the balanced distribution of vegetation elements and ensuring that the green space is aesthetically pleasing.

Planning of Paths and Walkways: The calculation of the perimeter is used to measure the length of paths, roads, and walkways in outdoor areas. This helps define the access route, the location of sidewalks, and the amount of materials needed for the construction of paths.

Design of Rest and Recreation Areas: In rest areas, playgrounds, and outdoor recreation areas, the calculation of the perimeter is essential for defining the boundaries of these areas. This allows for proper planning of play and rest spaces.

Delimitation of Ponds and Water Features: In projects that include ponds, fountains, or water features, the calculation of the perimeter is used to define the shape and size of these features. It is also relevant for determining the amount of lining or material needed for their construction.

Zoning of Outdoor Spaces: The calculation of the perimeter helps divide outdoor areas into functional zones, such as barbecue areas, rest zones, and gardens. This facilitates the organization and design of outdoor spaces.

Edge and Boundary Design: Landscape designers use the calculation of the perimeter to define the edges and boundaries of different areas. This includes creating garden edges, lawn borders, and areas with specific features.

Assessment of Available Space: The calculation of the perimeter helps assess how much space is available for planning gardens and landscaping elements. This is important to ensure that the design fits within the existing space.

In summary, the calculation of the perimeter is essential in landscaping projects, as it contributes to the planning and design of outdoor spaces that are visually appealing, functional, and appropriate for their purpose. It facilitates the distribution of vegetation elements, paths, and landscape features, ensuring that the outdoor space design is harmonious and meets the project's objectives.

Topography: In topography, the calculation of the perimeter is used to measure the length of land boundaries, plots, and geographical areas. It is also relevant for determining distances and areas on topographic maps.

The calculation of the perimeter plays a crucial role in topography, a discipline focused on the measurement and accurate representation of the Earth's surface. In topography, the calculation of the perimeter is applied in various situations.

Measurement of Property Boundaries: In topography, the calculation of the perimeter is used to measure the boundaries of properties, lands, plots, or specific geographical areas. This is fundamental for establishing ownership and clearly defining the dimensions of a property.

Determination of Distances: The calculation of the perimeter is used to determine distances between points on the Earth's surface. This is relevant in cartography and topography for measuring the length of roads, property boundaries, and other geographical elements.

Area Calculation: In addition to measuring distances, the calculation of the perimeter is essential for calculating areas of lands and regions on topographic maps. The perimeter defines the outline of an area, and this outline is used to determine the area of the enclosed space.

Delineation of Boundaries on Topographic Maps: Topographic maps often represent property boundaries, roads, and other geographical features. The calculation of the perimeter is applied to accurately determine the location of these boundaries on a map.

Land Planning and Development: The calculation of the perimeter is relevant in land planning and development, as it helps define the available areas and the boundaries of a construction or urban development project.

Monitoring Changes in the Land: In continuous-use topography applications, the calculation of the perimeter is used to monitor changes in the length of property boundaries or geographical areas over time.

Establishment of Project Boundaries: In civil engineering, urban planning, and construction projects, the calculation of the perimeter is applied to establish the boundaries of a project, which is important for design and execution.

Calculation of Road and Access Lengths: Measuring the perimeter of roads and access routes is relevant in transportation and road planning projects to determine the total length of road segments.

In summary, the calculation of the perimeter is an essential component of topography, as it allows measuring land boundaries, determining distances, and calculating geographical areas. These applications are fundamental in urban planning, cartography, construction, and many other areas related to the Earth's surface.

Material Estimation: In construction and manufacturing, the calculation of the perimeter is essential for estimating the amount of required materials. This is important to avoid waste and ensure that there are enough resources for a project.

Material estimation is a critical application of perimeter calculation in construction and manufacturing. By calculating the perimeter of a structure or component, the quantities of materials needed for a project can be accurately estimated.

Construction Materials: By measuring the perimeter of a structure, such as a wall, building, or fence, the amount of bricks, concrete blocks, wood, plaster, or other construction materials needed can be calculated. This ensures that enough materials are purchased or manufactured to complete the project without waste.

Fences and Barriers: In fencing and barrier projects, the calculation of the perimeter is essential to determine the total length of fencing required. This includes the number of posts, fence panels, and fastening materials needed.

Covers and Linings: For cladding or covering projects, such as wall panels, roofs, or pavements, the calculation of the perimeter is relevant to determine the number of panels, tiles, sheets, or lining material needed to cover a specific surface.

Ring-Shaped Structures: In components with a ring shape, such as pipes, ducts, or irrigation systems, the calculation of the perimeter is essential for estimating the length of these components. This helps acquire the correct amount of pipes or ducts without excess.

Wiring and Pipes: In electrical and plumbing projects, calculating the perimeter of an area or space allows estimating the length of wires, pipes, and conduits needed. This is fundamental to ensuring that there is enough material for installation.

Solar Panels: In solar energy projects, such as installing solar panels on roofs or land, the calculation of the perimeter is used to determine the number of panels required and the length of mounting structures.

Project Costs: Accurate material estimation based on perimeter calculation is essential for preparing precise budgets for construction and manufacturing projects. Material costs can be a significant part of the overall budget.

Waste Prevention: By adequately estimating the amount of materials, resource waste is avoided, and the need for restocking during the project is minimized. This saves time and money.

In summary, the calculation of the perimeter plays a fundamental role in material estimation in construction and manufacturing. It allows professionals and contractors to accurately determine how many materials are required, which is essential for efficiency and resource management in construction and manufacturing projects.

Design of Fences and Edges: In fencing and edging projects, the calculation of the perimeter is used to determine the amount of material needed, such as fences, railings, or garden borders.

The calculation of the perimeter is an essential part of fence and edging design projects. Determining the appropriate amount of material is fundamental to ensuring that the project is completed efficiently and that resources are used effectively.

Measurement of Boundaries: The calculation of the perimeter is used to measure the boundaries of the area to be fenced or edged. This is important to clearly define the area and ensure that the fence or edge fits properly within the available space.

Material Estimation: Once the perimeter has been calculated, the amount of materials needed can be estimated. This includes the number of posts, fence panels, railings, or edging material required to complete the project.

Design of the Fence: The calculation of the perimeter is also relevant for the design of the fence. It allows determining the total length of the fence and how the posts, panels, or other components will be distributed. This is crucial to ensure that the fence is built uniformly and attractively.

Costs and Budget: Accurate estimation of necessary materials based on perimeter calculation contributes to preparing an accurate budget for the project. This is essential to avoid unnecessary expenses and ensure that the project is carried out within the established financial limits.

Aesthetic Design: The calculation of the perimeter can also influence the aesthetic design of the project. It allows determining the height and length of the fence, as well as the placement of decorative elements such as ornamental posts or design details.

Access Planning: The calculation of the perimeter is used to plan the location of gates and entrances in a fence. This is essential to provide access to the area surrounding the fence.

Compliance with Regulations: In some cases, local regulations or zoning laws may establish specific requirements for fences. The calculation of the perimeter ensures that the fence complies with these requirements.

Security and Privacy: The calculation of the perimeter can also influence the security and privacy provided by a fence. It allows determining the appropriate height and placement of the fence to meet the homeowners' needs.

The calculation of the perimeter plays an integral role in fence and edging design projects, as it affects both the planning and execution of the project. Accurate estimation of the amount of material needed and proper design of the fence or edge are fundamental to achieving a satisfactory outcome and meeting the needs of the project and the homeowners.

Design of Routes and Roads: In road engineering, the calculation of the perimeter is relevant for measuring the length of roads, routes, and paths. This is important in planning and designing transportation systems.

The calculation of the perimeter plays a fundamental role in the design of routes and roads in the field of road engineering. Accurate measurement of the length of roads, routes, and paths is essential for planning, designing, and constructing efficient and safe transportation systems.

Measurement of Road Lengths: The calculation of the perimeter is used to measure the total length of planned roads, routes, and paths. This is essential to determine the distance that needs to be traveled from the starting point to the destination, which helps in planning routes and transportation systems.

Design of Curves and Turns: The calculation of the perimeter is applied to measure the length of curves, turns, and sections of roads that require specific design. This is important to ensure that curves are safe and comfortable for drivers.

Cost Estimation: By calculating the perimeter of a road or route, the amount of construction material, such as asphalt or concrete, needed to complete the project can be estimated. This is fundamental for preparing accurate budgets.

Planning of Signage: The measurement of the perimeter contributes to the planning of the location of traffic signs, road markings, and directional signs on the road. This is essential for road safety and guiding drivers.

Design of Intersections: In the design of road intersections, the calculation of the perimeter is used to define the length and shape of access and exit ramps, as well as the placement of turning lanes. This ensures that intersections are safe and functional.

Traffic Management: The measurement of the perimeter is relevant for planning and managing traffic on roads and routes. It helps determine speed limits, overtaking restrictions, and other traffic regulations.

Environmental Impact Assessment: The calculation of the perimeter is also important in assessing the environmental impact of road construction projects. It allows evaluating the scope of the project and its impact on the surrounding environment.

Planning of Toll Booths: For roads with toll systems, the calculation of the perimeter is relevant for determining the location and number of toll booths needed along the route.

Quality Control: The measurement of the perimeter is used in quality control during road construction to ensure that the length of the constructed road meets the design specifications.

In computational geometry, the calculation of the perimeter is a fundamental operation used in various algorithms and software to process and manipulate geometric figures. Some contexts in which perimeter calculation is relevant in computational geometry include:

Triangulation of Polygons: In polygon triangulation, which is a central problem in computational geometry, a polygon is divided into triangles. The calculation of the perimeter of the polygon is necessary to determine the lengths of the sides of these triangles.

Areas and Lengths: Computational geometry algorithms often require the calculation of areas and lengths. The perimeter of a figure is an essential part for calculating the lengths of the sides of a polygon or the circumference of a figure.

Comparison of Figures: In algorithms that involve the comparison of geometric figures, the calculation of the perimeter is used to determine which of two figures has a greater or lesser perimeter.

Selection of Reference Points: In generating reference points or nodes in geometric figures, perimeter calculation can be useful for evenly distributing these points along the figure's outline.

Optimization and Analysis: In the optimization and analysis of geometric figures, such as the placement of sensors in a wireless network or route planning for robots, perimeter calculation is relevant for minimizing distances or maximizing coverage of a region.

Image Processing: In image processing, perimeter calculation is used to delineate and analyze objects in digital images. It can help identify contours and features of interest in an image.

Mesh Generation: In the generation of three-dimensional meshes or surfaces, perimeter calculation is applied to define the connections between vertices and edges in the mesh.

Collision Detection: In simulation and video games, perimeter calculation is relevant for detecting collisions between objects and characters in a virtual environment.

The calculation of the perimeter plays an essential role in computational geometry and in a wide range of applications involving the processing and manipulation of geometric figures in the digital realm. It is used to measure, analyze, and make decisions related to geometric figures in diverse contexts, such as polygon triangulation, 3D mesh generation, and collision detection in video games.

It is an essential component in the design and planning of routes and roads. It ensures that roads are safe, functional, and efficient, and that the regulations and standards related to transportation are met. Additionally, it contributes to the accurate estimation of costs and resources needed for road infrastructure construction.

It plays a fundamental role in many areas, from planning and constructing structures to estimating materials and measuring land. It provides valuable information for decision-making and the efficient execution of projects across a wide variety of disciplines.

14.Algebra: Using letters (variables) to represent numbers in equations

Algebra is a branch of mathematics that uses letters (variables) and symbols to represent numbers and express mathematical relationships in the form of equations. Algebraic variables, often represented by letters such as "x," "y," "a," and "b," allow for the generalization and resolution of a wide range of mathematical problems.

Variables: Variables are symbols that represent unknown numbers or values that can vary. For example, in the equation "x + 5 = 10," "x" is a variable that represents an unknown number that must be determined. Variables are symbols used to represent unknown numbers or values that can vary in a mathematical equation or expression. These variables allow for the generalization and resolution of a wide range of mathematical and scientific problems. Unknowns: Variables are used to represent unknowns in equations or mathematical problems. When solving an equation, the goal is to find the specific value of the variable that makes the equation true. Notation: Variables are usually represented by letters, such as "x," "y," "a," "b," or any other letter of the alphabet. The choice of letter is generally arbitrary and is made for convenience.

Unknown Value: When working with an equation, the variable represents a value that is not yet known, and the goal is to determine that value through the resolution of the equation.

Flexibility: Variables can take different values in different contexts. For example, if you have the equation "x + 3 = 7," then "x" represents a value that, in this case, is equal to 4. However, in another equation, "x" could represent a different value.

Uses in Science and Mathematics: Variables are fundamental in formulating and solving problems in fields such as physics, chemistry, economics, statistics, and engineering. They are also used to model and understand phenomena and mathematical relationships in science and applied mathematics. Dependent and Independent Variables: In some equations and relationships, dependent and independent variables are used. The independent variable is considered as a value that is selected or controlled, while the dependent variable is the one that results from the independent variable.

Applications in Modeling: In science and mathematical modeling, variables are used to represent quantities that influence a phenomenon or process, allowing for the analysis and prediction of behaviors.

Variables are a fundamental part of algebra and mathematics in general. They provide a way to represent and solve problems in which a specific value is unknown, and that value is sought through mathematical calculations. Equations: Equations are mathematical expressions that establish an equality between two expressions. They usually contain variables and numbers.

Solving an equation involves finding the value or values of the variable that make the equality true. For example, in the equation "$2x - 3 = 7$," the goal is to find the value of "x" that satisfies the equation. Equations are mathematical expressions that establish an equality between two expressions. They often contain variables and numbers, and solving an equation involves finding the value or values of the variable that make the equality true.

Linear Equations: Linear equations are a common type of equation in which the variables are raised to the power of one, and the operations involve the addition and subtraction of terms. For example, "$2x - 3 = 7$" is a linear equation.

Solution of an Equation: The solution of an equation is the value or set of values that, when substituted into the equation, make the equality true. In the previous example, the solution would be "$x = 5$" since "$2 * 5 - 3$" equals 7.

Quadratic Equations: Quadratic equations are a type of equation in which the variable is raised to the second power, and the operations involve terms with exponent 2. For example, "$x^2 - 4x + 4 = 0$" is a quadratic equation.

Non-Linear Equations: Non-linear equations are those in which the variables are raised to powers other than one and may involve more complicated terms. Solving non-linear equations may require more advanced techniques. Systems of Equations: A system of equations is a set of two or more equations with multiple variables. The solution of a system of equations involves finding the values of the variables that satisfy all the equations in the system. Inequalities: Inequalities are mathematical expressions that establish relationships of inequality rather than equality. For example, "$2x < 10$" is an inequality that states that "$2x$" is less than 10.

Problem Solving: Equations are used to solve a wide variety of mathematical and scientific problems, from physics and economics to engineering and statistics. They provide a powerful tool for modeling and analyzing real-world situations. Equations are an essential part of mathematics and are used to represent mathematical relationships in a wide range of disciplines. Solving equations involves finding the values of the variables that make the equality (or inequality in the case of inequalities) true, which is fundamental for understanding and solving mathematical and scientific problems.

Algebraic Expressions: These are combinations of numbers, variables, and mathematical operators. For example, "3x + 2" is an algebraic expression in which "3x" is a term involving the variable "x," and "2" is a constant term. Algebraic expressions are combinations of numbers, variables, and mathematical operators. These expressions can take various forms and are used to represent mathematical relationships and perform mathematical calculations.

Terms: Algebraic expressions are composed of terms, which can be terms involving variables and constant terms. In your example, "3x" is a term involving the variable "x," and "2" is a constant term.

Coefficients: Coefficients are the numbers that multiply the variables in a term. In "3x," the coefficient is 3.

Operators: Algebraic expressions use mathematical operators such as addition, subtraction, multiplication, and division to combine terms and perform mathematical operations.

Operations: Algebraic expressions are used to perform mathematical operations such as simplification, expansion, factoring, and solving equations. Evaluation: Algebraic expressions can be evaluated by replacing the variables with specific values and performing the corresponding mathematical operations. For example, if "x = 2," then "3x + 2" would be evaluated as "3 * 2 + 2 = 8." Polynomials: Algebraic expressions can be polynomials, which are algebraic expressions with multiple terms. For example, "$2x^2 - 5x + 3$" is a polynomial. Simplification: Algebraic expressions are often simplified to reduce them to a more compact or manageable form.

Applications: Algebraic expressions are applied in a variety of fields, including physics, economics, statistics, engineering, and data science, to model and solve mathematical and scientific problems.

Algebraic expressions are a fundamental tool in mathematics and science, as they allow for the representation and analysis of a wide range of real-world situations and solving mathematical problems in various contexts. Operators: Mathematical operators, such as addition, subtraction, multiplication, and division, are used in algebra to perform operations on algebraic expressions. For example, "2x + 3" uses the addition operator "+".

Mathematical operators are symbols used to perform operations on algebraic expressions and equations. Each operator has a specific meaning and is applied to the numbers, variables, or terms in an algebraic expression. Addition (+): The addition operator is used to add two or more numbers or terms in an algebraic expression. For example, "2x + 3" implies the addition of "2x" and "3."

Subtraction (-): The subtraction operator is used to subtract one number or term from another in an algebraic expression. For example, "4 - x" implies subtracting "x" from "4."

*Multiplication (× or): The multiplication operator is used to multiply two or more numbers or terms. For example, "3x" implies multiplying "3" by "x." Division (÷ or /): The division operator is used to divide one number or term by another in an algebraic expression. For example, "6 ÷ 2" implies dividing "6" by "2."
Exponent (^): The exponent operator is used to raise a number or term to a specific power. For example, "x^2" means "x" raised to the second power. Square Root ($\sqrt{}$): The square root operator is used to calculate the square root or any other root of a number or term. For example, "$\sqrt{9}$" represents the square root of 9, which is 3.

Equal (=): Although it is not a mathematical operator in the traditional sense, the equal sign is used to establish equality between two expressions or terms. For example, "2x = 8" establishes that "2x" is equal to "8." Inequality (<, >, ≤, ≥): Inequality operators are used to express relationships of inequality between numbers or expressions. For example, "x < 5" indicates that "x" is less than 5.

These mathematical operators are fundamental for performing calculations and solving equations in algebra and mathematics in general. They are used to express mathematical relationships and represent a wide range of situations in science, engineering, and other disciplines.

Polynomials: These are algebraic expressions that consist of terms with coefficients and variables raised to integer exponents. Examples of polynomials are "3x^2 + 2x - 1" and "a^3 - 2a + 5."

Polynomials are algebraic expressions that consist of terms, and each term includes a multiplicative coefficient, a variable raised to an integer exponent, and optionally, a constant.

Terms: Polynomials are composed of terms. Each term in a polynomial is combination of a coefficient, a variable, and an exponent. For example, in the polynomial "3x^2 + 2x - 1," the terms are "3x^2," "2x," and "-1." Coefficients: Coefficients are numbers that multiply the variables in each term. In the polynomial "3x^2 + 2x - 1," the coefficients are 3, 2, and -1.

Variables: The variables in a polynomial represent unknown quantities or values that can vary. In the mentioned examples, "x" is the variable.

Exponents: Exponents are integers that indicate to what power the variable is raised in a term. For example, in "3x^2," the exponent is 2, which means that "x" is squared.

Degree of a Polynomial: The degree of a polynomial is the highest exponent among all the terms of the polynomial. For example, the degree of the polynomial "3x^2 + 2x - 1" is 2 since the term with the highest exponent is "3x^2."

Classification: Polynomials are classified according to their degree and number of terms. For example, a polynomial of degree 0 is a "monomial" (one term), a polynomial of degree 1 is a "binomial" (two terms), and a polynomial of degree 2 or more is a "trinomial" or a "general polynomial."

Operations: Various operations can be performed on polynomials, such as addition, subtraction, multiplication, and division. Simplification and factoring of polynomials are common tasks.

Applications: Polynomials are used in a wide variety of fields, from physics and engineering to economics and statistics, to model mathematical relationships in real-world situations.

Polynomials are a fundamental part of algebra and mathematics in general. They are used to represent and analyze mathematical relationships across a wide range of disciplines and to solve mathematical problems in various contexts. Systems of Equations: A system of equations is a set of two or more equations that share common variables. Solving a system of equations involves finding the values of the variables that satisfy all the equations simultaneously.

A system of equations is a set of two or more equations that involve the same variables and are used to model situations where the values of those variables are unknown. Solving a system of equations involves finding the values of the variables that make all the equations in the system true at the same time. Common Variables: In a system of equations, the equations share common variables. These variables represent unknown quantities that are being sought.

Solution of a System: The solution of a system of equations is a set of numerical values for the variables that satisfies all the equations in the system simultaneously. In other words, it is the set of values that makes all the equations true at the same time.

Types of Solutions: A system of equations can have different types of solutions, which include a unique solution (a single set of values that satisfies the system), infinitely many solutions (when all the equations are equivalent and represent the same line or plane), or no solution (when the equations are inconsistent and do not have a common set of values).

Methods of Resolution: There are several methods for solving systems of equations, such as the substitution method, elimination method, and matrix method. The choice of method depends on the nature of the system and the preferences of the solver.

Applications: Systems of equations are applied in a wide range of fields, from physics and economics to engineering and data science. They are used to model situations where multiple factors interact and relate through equations. Intersection of Lines or Planes: In geometry, solving a system of equations in two dimensions is equivalent to finding the intersection point of two lines, while in three dimensions, it is equivalent to finding the intersection point of two planes.

Notation: Systems of equations are commonly represented with notation such as "$x + y = 5$" and "$2x - 3y = 10$," where "x" and "y" are the shared variables and the equations are written one below the other.

Solving systems of equations is an essential mathematical skill used to address a variety of problems in mathematics and applied fields. It can provide solutions to complex problems in science, engineering, economics, and many other disciplines.

Inequalities: Inequalities are mathematical expressions that express relationships of inequality rather than equality. For example, "$2x > 8$" is an inequality that states that "$2x$" is greater than 8.

Inequalities are mathematical expressions that establish relationships of inequality instead of equality. They are used to represent situations in which one quantity is greater than, less than, or different from another quantity. Inequality Symbols: Inequalities include inequality symbols, such as "$<$" (less than), "$>$" (greater than), "$\leq$" (less than or equal to), and "$\geq$" (greater than or equal to). These symbols indicate the relationship between the two expressions in the inequality.

Variables: Like equations, inequalities can involve variables, which represent unknown quantities or values that can vary. For example, in the inequality "$2x > 8$," "x" is the variable.

Solution of an Inequality: The solution of an inequality is a set of values for the variable that makes the inequality true. For example, for the inequality "$2x > 8$," a solution would be "$x > 4$" because any value of "x" greater than 4 makes the inequality true.

Intervals: Solutions to inequalities are often represented in the form of intervals on the number line. For example, "x > 4" is represented as an open interval starting from 4 onward on the number line.

Combination of Inequalities: Multiple inequalities can be combined using logical operators such as "and" (conjunction) and "or" (disjunction) to express more complex relationships.

Applications: Inequalities are applied in a variety of fields, from economics and physics to linear programming and probability theory. They are used to model constraints and limitations in real-world problems.

Resolution of Inequalities: Solving inequalities involves finding the set of values that satisfy the inequality. This often involves algebraic manipulation and graphical representation on the number line.

Inequalities are a fundamental tool in mathematics and are used to describe a wide range of situations where relationships of inequality are relevant. They are especially important in the realm of optimization and decision-making in problem-solving.

Factoring: It is the process of breaking down an algebraic expression into simpler factors. Factoring is useful for simplifying equations and expressions. Factoring involves breaking down an algebraic expression into simpler factors, which can facilitate the simplification of equations and expressions. Goal of Factoring: The goal of factoring is to express an algebraic expression in an equivalent but more manageable form. By breaking down an expression into simpler factors, calculations can be simplified, and equations can be solved more efficiently.

Factors: Factors are the simplest expressions into which an algebraic expression is decomposed. These factors can be monomials, binomials, trinomials, or other algebraic expressions.

Factoring Methods: There are several factoring methods, such as factoring by common factor, factoring by difference of squares, factoring by perfect square trinomial, factoring by non-perfect square trinomial, factoring by grouping, and factoring by decomposition into prime factors.

Examples of Factoring: Here are some examples of factoring:Factoring by common factor: $2x + 4$ factors to $2(x + 2)$.

Factoring by difference of squares: $x^2 - 4$ factors to $(x + 2)(x - 2)$. Factoring by perfect square trinomial: $x^2 + 4x + 4$ factors to $(x + 2)^2$.

Applications of Factoring: Factoring is applied in a variety of fields, from algebra and calculus to number theory and statistics. It is used to simplify equations, solve systems of equations, find roots of polynomials, and simplify algebraic fractions.

Simplification of Equations: Factoring is a valuable technique for simplifying equations as it can lead to the cancellation of common terms on both sides of the equation.

Problem Solving: Factoring is an essential tool in mathematical problem-solving and in representing algebraic relationships in a more manageable way. Factoring is a fundamental and versatile mathematical skill used in various contexts in mathematics and related disciplines. It helps to simplify and understand algebraic expressions more effectively.

Evaluation: In algebra, algebraic expressions are evaluated by replacing the variables with specific values and performing the corresponding mathematical operations. For example, if "$x = 3$," then "$2x + 5$" would be evaluated as "$2 * 3 + 5 = 11$."

Evaluation involves replacing the variables in an algebraic expression with specific values and performing the corresponding mathematical operations. Substitution of Variables: In evaluation, the variables in an algebraic expression are replaced by known numerical values. These values are known as "assignments" or "substitutions."

Mathematical Operations: Once substitutions have been made, the specified mathematical operations in the algebraic expression are applied. These operations may include addition, subtraction, multiplication, division, exponentiation, among others.

Result of the Evaluation: The result of the evaluation is a numerical value that represents the value of the algebraic expression after replacing the variables and performing the operations. In the provided example, if "$x = 3$," then "$2x + 5$" evaluates to "$2 * 3 + 5 = 11$."

Variables and Constants: In an algebraic expression, variables can represent unknown values or values that can vary. Constants are fixed numerical values. Evaluation is used to determine the value of the expression when the values of the variables and constants are known.

Applications: Evaluation is a fundamental tool in algebra and is used in a variety of contexts. It is applied in solving equations, simplifying algebraic expressions, and interpreting mathematical results in terms of specific quantities. Evaluation is essential in mathematics and science, as it allows for the assignment of numerical values to variables and expressions, facilitating the understanding and application of mathematical concepts in concrete situations.

15.Equations: Mathematical expressions that show equality, like 2x + 3 = 7

Equations are mathematical expressions that show equality between two expressions or quantities. Here are some key points about equations:

Equality: An equation establishes that two expressions are equal. It uses an equal sign ("=") to indicate that the left side of the equation is equal to the right side. The equal sign ("=") is essential in equations, as it indicates that two expressions or quantities are equal.

Balance: The equality in an equation represents a mathematical balance. It means that the values on both sides of the equation are equivalent and that both sides are balanced.

Left Side and Right Side: In an equation, the left side contains one mathematical expression, and the right side contains another expression. The equal sign is used to indicate that both expressions are equivalent and have the same value.

Reflexive Property: Equality is a reflexive property. This means that any number or expression is equal to itself. For example, "3 = 3" is a true equation.

Examples of Equality: Here are additional examples of equations that represent equality:

"2x = 10" indicates that "2x" is equal to 10.

"a + 5 = 9" indicates that "a + 5" is equal to 9.

"4y - 7 = 5" indicates that "4y - 7" is equal to 5.

Solving Equations: Solving an equation involves finding the values of the variables that make the equality true. This often involves performing mathematical operations on both sides of the equation to isolate the variable.

The equality in equations is a fundamental concept in mathematics and is used to establish relationships and solve mathematical and scientific problems. It allows modeling situations in which quantities are equivalent and remain balanced.

Variables: Equations often include variables, which are symbols that represent unknown quantities or values that can vary. In your example, "x" is the variable. Variables are symbols that represent unknown quantities or values that can vary.

Representation of Unknown Quantities: Variables are used in equations to represent quantities that are not known with certainty or that may vary in different situations. For example, in the equation "$2x + 3 = 7$," the variable "x" represents an unknown quantity that is being sought.

Notation of Variables: Variables are commonly expressed as letters of the alphabet, such as "x," "y," "a," "b," "c," etc. It is also common to use subscripts to distinguish between different related variables.

Solutions: Solving an equation involves finding the values of the variables that make the equality true. These values are called "solutions" of the equation. In the previous example, "$x = 2$" is a solution because it satisfies the equation.

Independent and Dependent Variables: In some contexts, such as in systems of equations, one variable may depend on another. For example, in a system of linear equations, "x" and "y" may be interdependent variables.

Applications in Science and Mathematics: Variables are used to model a wide variety of phenomena in science and mathematics. For example, in physics, "t" may represent time, and in economics, "p" may represent price.

Problem Solving: Equations with variables are used to solve problems in mathematics and science, as well as in real-world situations where unknown values need to be found. Variables are a fundamental part of equations and are essential for describing mathematical relationships and solving problems that involve unknowns or variable quantities.

Constants: In addition to variables, equations may contain constants, which are fixed numerical values. In your equation, "2," "3," and "7" are constants. Constants are fixed numerical values used in equations to represent known quantities or values that do not change in a given situation.

Fixed Values: Constants are specific numbers that do not vary in the context of the equation. For example, in the equation "$2x + 3 = 7$," the numbers "2," "3," and "7" are constants.

Role in Operations: Constants are used in mathematical operations along with variables to form algebraic expressions. In the example, "$2x$" is an expression that includes a constant ("2") multiplied by a variable ("x").

Coefficients: In linear equations, the constants that multiply the variables are called "coefficients." In "$2x$," the coefficient of "x" is 2.

Application in Problems: Constants are used to model known values in mathematical and scientific situations. For example, if solving a physics problem involving the speed of an object, the initial speed might be a known constant.

Equations with Constants: Equations can include both variables and constants. Constants often represent values that are known or established beforehand in a problem, while variables represent unknown or varying quantities.

Simplification of Expressions: In the simplification of equations or algebraic expressions, constants can be combined or grouped to reduce the expression to a simpler form. Constants are an essential part of equations and are used to describe mathematical relationships and model real-world situations where some quantities remain fixed or are known. This is fundamental for problem-solving in mathematics and sciences.

Resolution: Solving an equation involves finding the value or values of the variable that make the equality true. The value that satisfies the equation is called a "solution."

The goal of solving an equation is to find the value or values of the variable that make the equality in the equation true. The value that satisfies the equation is called a "solution."

Solution: A solution of an equation is a value or set of values that, when substituted into the equation, makes the equality true. For example, in the equation "$2x + 3 = 7$," the solution is "$x = 2$" because when "x" is substituted by 2, the equation is true: "$2 * 2 + 3 = 7$."

Goal of Resolution: The goal when solving an equation is to find which values of the variable make the equation true. This involves isolating the variable on one side of the equation.

Mathematical Operations: Solving an equation often involves performing a series of mathematical operations on both sides of the equation to isolate the variable. These operations may include addition, subtraction, multiplication, division, and others.

Verification: After finding a solution, it is important to verify that the proposed value actually satisfies the equation. This involves substituting the value into the original equation and checking that both sides are equal.

Multiple Solutions: Some equations have more than one solution, while others may have no solution at all. It will depend on the nature of the equation and its coefficients.

Applications: The resolution of equations is used in a wide variety of fields, from pure mathematics to physics, economics, and engineering. It is applied to model real-world situations and make decisions based on quantitative data. The resolution of equations is a fundamental process in mathematics and science, as it allows finding answers to mathematical problems and modeling situations where quantities are unknown but can be calculated.

Operations: Equations involve mathematical operations, such as addition, subtraction, multiplication, division, and exponentiation. These operations are applied to both the left side and the right side of the equation.

Equations involve mathematical operations applied to both the left side and the right side of the equation. The mathematical operations used in equations include addition, subtraction, multiplication, division, exponentiation, and others.

Addition (+): The addition operation is used to add values. In an equation, terms can be added or subtracted on both sides to balance the equation.

Subtraction (-): The subtraction operation is used to subtract values. As with addition, one can subtract or add on both sides of the equation.

Multiplication (×): The multiplication operation is used to increase or decrease values. Multiplying or dividing terms by a factor on both sides of the equation allows changing the value of the variable.

Division (÷): The division operation is used to divide values. Dividing or multiplying terms on both sides of the equation is a common way to isolate the variable.

Exponentiation (^): The exponentiation operation is used to raise values to a power. Raising or taking roots on both sides of the equation can change the value of the variable.

Combined Operations: In more complex equations, multiple operations may be used together. For example, an equation could involve both additions and multiplications.

Maintaining Balance: Operations are used to maintain balance in the equation. Any operation performed on one side of the equation must have an equivalent effect on the other side to keep the equality intact.

Order of Operations: When solving equations, it is important to follow the order of mathematical operations. This ensures that operations are performed correctly and that the equation is solved accurately.

Mathematical operations are an essential part of solving equations and are used to isolate the variable and find solutions. The goal is to find values of the variable that make the equality in the equation true.

Solution of Equations: An equation can have one or multiple solutions. For example, in the equation "2x + 3 = 7," the solution is "x = 2" because when "x" is substituted by 2, the equation is true: "2 * 2 + 3 = 7."

An equation can have one or multiple solutions. The solution of an equation is the value or set of values of the variable that makes the equality in the equation true.

Unique Solution: Some equations have a unique solution, meaning there is only one value of the variable that satisfies the equation. In the example you provided, "2x + 3 = 7," the solution is unique and is "x = 2."

Infinite Solutions: Some equations have infinite solutions. This occurs when any value of the variable satisfies the equation. For example, the equation "3x = 3" has infinite solutions since any value of "x" that makes "3x" equal to 3 is a solution.

No Solution: Some equations have no solution. This happens when there is no value of the variable that satisfies the equation. For example, the equation "2x + 3 = 1" has no solution since there is no value of "x" that makes "2x + 3" equal to 1.

Verification: It is important to verify that a proposed solution satisfies the equation. This involves substituting the value of the variable in the original equation and checking that both parts are equal. Verification ensures that the solution is correct.

General Expression: When solving equations, one often seeks to find the general expression that describes all possible solutions rather than a specific solution.

Degree of the Equation: The degree of an equation refers to the highest exponent in the variable. First-degree (linear) equations have a single solution, while higher-degree equations may have multiple solutions.

The ability to find solutions for equations is fundamental in mathematics and is applied in various disciplines, from physics to economics. The resolution of equations allows modeling and solving problems where quantities are unknown or variable.

Applications: Equations are used in a wide variety of fields, from physics and engineering to economics and data science. They are applied to model mathematical relationships and solve problems in real-world situations. Equations have a wide range of applications in various fields.

Physics: In physics, equations describe mathematical relationships between variables such as speed, acceleration, force, and energy. Newton's motion equations and the equation of the law of universal gravitation are examples.

Engineering: Engineers use equations to design and analyze systems and structures. For example, in civil engineering, equations are used to calculate the strength of materials in bridges and buildings.

Economics: In economics, equations are applied to model economic relationships, such as supply and demand, production costs, and inflation. These equations are fundamental in economic and financial decision-making.

Data Science: In data science, equations are used to model and analyze data. Regression equations, for example, are used to predict relationships between variables in datasets.

Biology: Equations are applied in biology to model populations, growth rates, and biochemical processes. For example, enzyme kinetics equations describe chemical reactions in biological systems.

Chemistry: Chemical equations are used to represent chemical reactions, helping to understand how elements and compounds interact and transform.

Medicine: In medicine, equations are applied in areas like pharmacokinetics to determine the appropriate dosage of medications based on patient metabolism.

Astronomy: Equations describe planetary motions, comet orbits, and other astronomical phenomena. Kepler's laws and Newton's law of gravitation are examples.

Technology: In technology, equations are used in the design and optimization of electronic systems, computer algorithms, and electrical circuits.

Education: Equations are taught in mathematics and are used to solve problems and teach fundamental mathematical skills.

Equations are a powerful tool for modeling mathematical relationships in a wide variety of fields and for addressing problems in real-world situations.

These applications demonstrate the importance of equations in science, technology, engineering, and mathematics (STEM) as well as in many other areas. Equations are an essential tool in mathematics and science, as they allow for the representation and resolution of problems, establishing relationships between quantities and making decisions based on quantitative data.

16.Inequalities: Expressions that show "greater than" or "less than" relationships instead of equality

Inequalities are mathematical expressions that establish relationships of inequality instead of equality. In an inequality, one of the following symbols is used: "<" (less than): Indicates that one value is less than another. For example, "x < 5" means that "x" is less than 5. ">" (greater than): Indicates that one value is greater than another. For example, "y > 3" means that "y" is greater than 3. "<=" (less than or equal to): Indicates that one value is less than or equal to another. For example, "z <= 8" means that "z" is less than or equal to 8. ">=" (greater than or equal to): Indicates that one value is greater than or equal to another. For example, "w >= 10" means that "w" is greater than or equal to 10.

Inequalities are used to represent a wide range of relationships in mathematics and are applied in situations where quantities are not necessarily equal but are compared in terms of size or magnitude. Inequalities are especially useful for describing restrictions and conditions in mathematical problems and real-world applications, such as budget planning, resource optimization, and decision-making.

Here are some additional applications of inequalities in various contexts:

Economics: In economics, inequalities are used to model budget constraints, resource limitations, and supply and demand conditions. For example, an inequality can describe the maximum quantity of a product that a company can produce given its resource limitations.

Inequalities are a fundamental tool for modeling and analyzing a variety of economic and financial situations. Here are some more specific applications in economics:

Budget Constraints: Inequalities are used to represent the budget constraints of individuals, families, or companies. For instance, a family may use an inequality to determine how much they can spend on food and housing given their monthly income.

Production Optimization: In the production of goods and services, inequalities are applied to model resource limitations, such as the availability of raw materials or production capacity. This is important for maximizing efficiency and profitability.

Supply and Demand: In supply and demand analysis, inequalities are used to represent restrictions on the quantity of a product that is available or desired for purchase. These inequalities help determine equilibrium prices and production quantities.

Investment Planning: In financial investments, inequalities are applied to model investment strategies and risks. For example, an inequality can describe risk constraints in an investment portfolio.

Natural Resource Management: Inequalities are used to set limits on the exploitation of natural resources, such as fishing, mining, and logging, to ensure long-term sustainability.

Taxes and Taxation: Inequalities are applied in tax planning and taxation to model the budget constraints of individuals and companies and to determine the optimal tax burden.

Distribution of Educational Resources: In education, inequalities are used to establish criteria for the allocation of educational resources, such as school funding, scholarships, and access to educational programs.

Monetary Policy: In formulating economic policies, inequalities are applied to model restrictions and targets, such as the target inflation rate and interest rates.

In summary, inequalities are an essential tool in economics for modeling and solving problems related to budget constraints, limited resources, and market conditions. They help economists and decision-makers make informed choices in a variety of economic and financial contexts.

Optimization: Inequalities are essential in optimization problems, where the goal is to find the best solution under certain constraints. For example, in linear programming, inequalities are used to maximize or minimize an objective function subject to constraints.

Inequalities play a crucial role in optimization problems, where the aim is to find the best possible solution within certain limitations or constraints. One of the most common approaches to optimization with inequalities is linear programming.

Linear Programming: Linear programming is an optimization method used in a wide variety of applications, from supply chain management to decision-making in businesses and organizations. In this approach, optimal values for decision variables are sought while satisfying linear constraints expressed as inequalities.

Objective Function: In a linear programming problem, an objective function is defined that is to be maximized or minimized. This function can represent profit, cost, time, or any other quantity to be optimized.

Constraints: Constraints are expressed as inequalities that limit the values that decision variables can take. These inequalities can represent resource constraints, capacity limitations, or any other type of limitation.

Optimal Solution: The optimal solution is the set of values for the decision variables that maximizes or minimizes the objective function while simultaneously satisfying all constraints. This solution provides the best combination of decisions given the restrictions.

Solution Algorithms: Specific algorithms are used to solve linear programming problems and find the optimal solution. Some of the most well-known methods include the simplex method and gradient method.

Applications: Linear programming is applied in a wide range of situations, such as production planning, resource allocation, product distribution, route planning, and decision-making in business management.

In summary, inequalities are essential in optimization problems, and linear programming is a widely used mathematical approach to address these problems. It allows for informed decision-making to maximize benefits or minimize costs while respecting constraints and limitations in business and management situations.

Environmental Science: In the management of natural resources and environmental conservation, inequalities are applied to set limits on resource exploitation, such as sustainable fishing and the management of protected areas.

In environmental science and natural resource management, inequalities play a fundamental role in formulating policies and strategies for the conservation and sustainable use of natural resources.

Sustainable Fishing: Inequalities are used to establish fishing quotas that limit the number of fish that can be caught in an effort to avoid overfishing and ensure the regeneration of fish populations.

Management of Protected Areas: Inequalities are applied in the management of national parks and nature reserves to establish limits on human activity, such as the construction of infrastructure or resource harvesting, in order to preserve biodiversity and ecosystems.

Species Conservation: Inequalities are used to establish restrictions on the hunting and capture of endangered or at-risk species, with the goal of protecting these species and their habitats.

Land Use Planning: In urban and rural planning, inequalities are applied to limit development in environmentally sensitive areas, such as flood zones or critical habitats.

Forest Resource Management: Inequalities are used in logging and sustainable forestry to set limits on the amount of timber that can be harvested without depleting forest resources.

Water Management: In the management of river basins and aquifers, inequalities are applied to set restrictions on water extraction to ensure a sustainable water supply.

Emission Restrictions: Inequalities are used in regulating pollutant emissions to limit the amount of contaminants released by industrial and other sources, in order to protect air and water quality.

Natural Reserve Planning: Inequalities are applied in identifying and selecting areas for the establishment of new nature reserves, considering factors such as habitat connectivity and protection of endangered species.

Inequalities play an essential role in decision-making related to environmental conservation and natural resource management, helping to establish limits and restrictions that are fundamental for the long-term sustainability of ecosystems and biodiversity.

Urban Design and Planning: In architecture and urban planning, inequalities are used to establish zoning regulations and building restrictions in urban areas.

In the field of architecture and urban planning, inequalities are employed to establish regulations and restrictions governing the development and zoning of urban areas.

Zoning: Inequalities are used to define zones in a city or urban area, determining what types of structures or activities are permitted in each zone. For example, restrictions can be established on building height, population density, land use (residential, commercial, industrial), and other parameters.

Height Regulation: Inequalities are applied to limit the height of buildings in specific areas of a city. This is crucial for maintaining a coherent urban aesthetic and avoiding excessive shadows in residential areas.

Land Use Planning: Inequalities help regulate land use in urban areas. For instance, restrictions may be set on the amount of land that can be allocated for construction and the amount of space that must be reserved for green spaces and parks.

Density Limitations: Inequalities are essential for controlling population density in urban areas. This may involve restrictions on the number of housing units allowed per area or occupancy restrictions.

Safety Distances: Inequalities are applied to establish safety distances between buildings, industrial facilities, and residential or commercial areas. This is important for the safety and well-being of residents and workers.

Parking Regulation: Inequalities are used to establish parking requirements for new developments, such as shopping centers or residential complexes. This influences the amount of space allocated for parking.

Historical Preservation: In areas with historic buildings or sites, inequalities may limit modifications and renovations to preserve historical character.

Infrastructure Planning: Inequalities also apply in infrastructure planning, such as the construction of roads, bridges, and utilities, and can set restrictions for the location and design of these structures.

Public Space Design: Inequalities may influence the planning and design of public spaces, such as parks, plazas, and pedestrian walkways.

The application of inequalities in architecture and urban planning is fundamental to ensuring the orderly and sustainable development of urban areas, balancing the needs for growth with the conservation of the environment, quality of life, and the safety of residents.

Health: In medical research and public health, inequalities can describe risk thresholds, such as safe levels of exposure to toxic substances.

Inequalities play an important role in medical research and public health by establishing risk thresholds and safe limits.

Toxicology and Chemical Exposure: In toxicology, inequalities are used to describe safe levels of exposure to chemical substances, such as industrial chemicals or medications. These inequalities establish thresholds above which exposure is considered dangerous.

Epidemiology: In epidemiology, inequalities can be used to model the spread of infectious diseases and establish critical immunization or control measure thresholds. For example, an inequality could describe the relationship between vaccination rates and the spread of a disease.

Nutrition and Diet: Inequalities are also applied in nutrition and public health to define guidelines for safe and healthy food consumption. They can set restrictions on daily intake of certain nutrients or harmful substances.

Water and Air Quality: In monitoring water and air quality, inequalities are used to set maximum allowable limits for pollutants and chemicals, ensuring that exposure levels are safe for human health.

Drug and Medication Research: Inequalities are useful in drug research for establishing safe and effective dosages. They are applied to define the relationship between the administered dose and side effects.

Healthcare Resource Management: In managing healthcare resources, inequalities can be used to establish admission or treatment thresholds based on hospital capacity and resource availability.

Disaster Response Planning: Inequalities are applied in disaster response planning, such as hospital capacity planning, establishing critical thresholds for activating additional resources in emergency situations.

In medical research and public health, inequalities contribute to establishing guidelines, regulations, and policies that protect public health and ensure safety in environments where exposure to risks may be harmful.

Education: In the field of education, inequalities are applied in defining criteria for passing and grading, as well as in the distribution of educational resources.

Passing Criteria: In educational systems, inequalities are used to establish passing criteria for exams and assessments. For example, an inequality might indicate that a student needs to achieve a score equal to or greater than a certain value to pass an exam.

Resource Distribution: Inequalities may be applied in the allocation of educational resources, such as scholarships and grants. For instance, academic scholarships may be subject to inequalities relating academic performance to eligibility for financial aid.

Inclusion Policies: Inequalities can also be used in designing inclusion and access policies in education. For example, an inequality could define admission requirements for special education programs.

Performance Evaluation: In evaluating student performance, inequalities can set thresholds for grade promotion, grading allocation, or participation in advanced education programs.

Resource Planning: Inequalities can be applied in planning educational resources, such as the allocation of classrooms and teachers based on capacity and demand.

Course Optimization: In educational institutions, inequalities can be used to optimize course scheduling, ensuring that time and capacity constraints are met.

Safety and Health: Inequalities may relate to the safety and health of students, establishing restrictions to ensure a safe and healthy educational environment.

In summary, inequalities are valuable tools in the field of education for establishing standards, criteria, and policies that influence decision-making and resource management in educational institutions.

Transportation: In transportation planning and logistics, inequalities are used to model capacity constraints on roads, airports, and public transportation networks.

Inequalities play a fundamental role in transportation planning and logistics, where they are used to model capacity constraints and make decisions related to the efficient management of transportation infrastructure.

Traffic Management: In traffic control, inequalities are used to model the capacity of roads and transit routes. For example, they can describe the relationship between traffic volume and speed, which is essential for congestion management and mobility planning.

Transportation Resource Allocation: Inequalities can be applied in the allocation of transportation resources, such as train schedules, flights, and public transport routes. This helps ensure that resources are used efficiently and within capacity constraints.

Route Optimization: In logistics and freight transportation, inequalities are used to optimize delivery routes and shipping schedules, considering time and capacity limitations on roads and ports.

Infrastructure Planning: Inequalities are useful in the planning and design of transportation infrastructure, such as bridges and airports. They help ensure that structures are safe and meet capacity standards.

Fleet Planning: Inequalities apply in fleet vehicle planning, ensuring that the number of vehicles and their capacity align with transportation demand without exceeding route and terminal capacity limits.

Public Transport Planning: In planning public transportation networks, inequalities are used to ensure that the schedules and capacity of services meet user needs.

Toll System Design: Inequalities are applied in the management of toll systems on roads and bridges, ensuring that fees align with capacity and demand constraints.

Overall, inequalities in transportation planning are essential to ensuring an efficient flow of people and goods, minimizing congestion, and ensuring safety in transportation infrastructure.

Security: In security, inequalities can describe conditions for the acceptance or rejection of certain actions or decisions.

In the field of security, inequalities are used to establish restrictions and conditions that influence decision-making and risk management.

Access Control: Inequalities can describe the conditions under which access to secure areas is allowed or denied. For instance, inequalities may define authorization thresholds based on credentials, such as access cards or identification codes.

Risk Management: In security risk management, inequalities are used to set limits on exposure to risks. For example, they can establish exposure thresholds to hazardous chemicals in a workplace environment.

Evacuation Planning: Inequalities may describe restrictions in the planning of evacuations from buildings or areas during emergencies. This could include capacity restrictions on emergency exits or maximum evacuation times.

Information Technology Security: Inequalities may apply in access control to computer systems and networks, establishing conditions for user authorization and data protection.

Transportation Security: In the transport of hazardous goods, inequalities are used to establish quantity restrictions and transportation conditions that minimize safety risks.

Construction Safety: Inequalities may apply in the construction of structures, establishing load and support capacity restrictions to ensure the safety of buildings and bridges.

Personal Data Protection: In the realm of privacy and data protection, inequalities may define conditions under which access to personal information is allowed, ensuring confidentiality and compliance with privacy regulations.

Inequalities play an important role in security by establishing limits and conditions that help mitigate risks, control access, and ensure safety in various situations and environments. They are a versatile mathematical tool that is applied in a variety of situations to represent relationships of inequality and establish limits in practical problems and data-driven decisions. Their ability to model restrictions and conditions makes them essential for informed decision-making across a wide range of fields.

17.Ratios: Comparing two quantities, like 2:5

Ratios are a way to compare two quantities or values through an expression that shows the relationship between them. Ratios are expressed as fractions or by two numbers separated by a colon or a colon and a slash. For example, the ratio "2:5" or "2/5" indicates the comparison between two quantities, in this case, the number 2 in relation to the number 5. Ratios can have various uses and applications in mathematics and problem-solving across different disciplines. Here are some examples of their application:

Proportions: Ratios are commonly used in proportions to compare quantities. For example, if the number of men (2) is compared to the number of women (5) in a group, the ratio 2:5 indicates that there are 2 men for every 5 women in that group. They are fundamental for expressing proportions and comparing quantities in a variety of situations. In your example, the ratio 2:5 represents the proportion of men to women in a group. This means that for every 2 men, there are 5 women in that group.

Proportions are an effective way to communicate the relative relationship between two sets of data, whether in the context of populations, samples, quantities, or any other type of quantitative comparison. They can be expressed as fractions or with a colon (such as 2:5) and are used in mathematics, statistics, social sciences, and many other fields to describe quantitative relationships. Proportions and ratios are important tools in data analysis and decision-making based on data.

Scales: In maps and plans, ratios are used to represent scale relationships. For example, a scale of 1:1000 on a map means that a distance on the map is one thousand times smaller than the actual distance on the ground. Scales are fundamental in cartography and the graphical representation of geography. When a ratio such as 1:1000 is used on a map, a scale relationship is being established between the distances represented on the map and the actual distances on the ground. In your example, a scale of 1:1000 means that one unit of length on the map is equivalent to 1/1000 of the same unit of length on the real ground. This implies that the distances on the map are reduced by a factor of 1000 compared to the actual distances.

Scales are essential for representations such as maps and plans to be useful and accurate. They allow people to correctly interpret distances and dimensions in relation to the real world. Scales are used in cartography, architectural design, engineering, surveying, and various applications where an accurate graphical representation of spatial information is required.

Scales can vary depending on the situation and the level of detail needed in the graphical representation.

Probability: In statistics and probability, ratios are used to express the probability of one event in relation to another. For example, if the probability of event A occurring is 2 times greater than the probability of event B occurring, the odds ratio would be 2:1. In statistics and probability, odds ratios are a common way to express the relative probability of one event occurring compared to another. The odds ratio is calculated by dividing the probability of event A occurring by the probability of event B occurring. If the probability of event A occurring is 2 times greater than the probability of event B occurring, the odds ratio would be 2:1.

Odds ratios are useful in making decisions based on probabilities and assessing risks. They are used in a wide range of applications, from medical research and epidemiology to risk assessment in insurance and finance. They are also fundamental in game theory and in interpreting results from experiments and statistical studies.

Speed and Distance: In problems related to motion and speed, ratios are used to compare speeds or distances. For example, if car A travels at a speed of 60 km/h and car B travels at 120 km/h, the speed ratio would be 60:120 or 1:2. Ratios are used in problems related to motion and speed to compare speeds or distances. In your example, if car A travels at 60 km/h and car B travels at 120 km/h, the speed ratio between them would be 60:120 or, simplifying, 1:2. This means that car B is traveling at a speed that is twice that of car A.

Speed and distance ratios are common in kinematics and dynamics problems and are used to understand and quantify the motion of objects in relation to other objects or references. They are also relevant in areas such as navigation, aviation, and traffic engineering, where decisions must be made based on the speed and distance traveled by vehicles or moving objects.

Finance: In finance, ratios are used to compare financial indicators, such as the price-earnings ratio (P/E ratio) in stocks, where the price of a stock is compared to its earnings. In the world of finance, ratios are important tools for evaluating and comparing financial indicators. The price-earnings ratio (P/E ratio) is a common example of a financial ratio. It is calculated by dividing the price of a stock by its earnings per share (EPS).

The P/E ratio is used to evaluate the valuation of a stock in the stock market. A high P/E ratio may indicate that investors are willing to pay a higher price for each dollar of the company's earnings, which could suggest an expectation of future growth. On the other hand, a low P/E ratio may indicate that the stocks are valued lower in relation to their earnings, which could be a sign that the stock is undervalued.

Financial ratios are also applied in assessing profitability, solvency, and efficiency of companies, and are fundamental in making investment and financing decisions.

Ratios are useful tools for expressing comparisons and relationships between quantities or values, and they are widely used in mathematics and practical problem-solving across various fields. Ratios are fundamental tools in mathematics and practical problem-solving in various fields. They allow for the expression of comparisons and relationships between quantities or values, facilitating the understanding and analysis of complex situations. Whether in mathematics, science, finance, engineering, statistics, or any other discipline, ratios are an effective way to quantify and communicate quantitative relationships.

Their versatility and applicability in a wide range of situations make ratios a valuable tool both in the academic and practical world. They can be used to make informed decisions, draw meaningful comparisons, and solve problems involving related quantities or values.

18.Proportions: Equal relationships between ratios

Proportions are relationships of equality between ratios. In other words, a proportion is an equality between two ratios, and it is generally expressed in the form "a:b = c:d," where "a" and "b" are numbers (or ratios) and "c" and "d" are different numbers (or ratios). This means that the first ratio "a:b" is equal to the second ratio "c:d."

Proportions are used to compare two sets of quantities in such a way that the relationship between the quantities of one set is equal to the relationship between the quantities of the other set. Proportions are common in mathematics, science, economics, and many other disciplines, and they are applied to solve problems involving proportional relationships.

For example, in a proportion problem, you might have a situation where "a" is the amount of money you earn in 5 hours, "b" is the amount of money you earn in 8 hours, "c" is the amount of money another person earns in 5 hours, and "d" is the amount of money the same person earns in 8 hours. If a:b = c:d, then you are expressing a relationship of proportional equality in terms of earnings per hour.

Proportions are a fundamental tool in mathematics and are applied in a wide variety of disciplines and situations to solve problems involving proportional relationships. Here are some areas where proportions are common and essential:

Mathematics: Proportions are used to solve problems of direct or inverse proportionality, such as those related to ratios, rates, and percentages. They are also fundamental in solving algebra and geometry problems.

Proportions are fundamental in mathematics and have applications in various concepts and areas:

Direct Proportionality: When two quantities are directly proportional, it means they increase or decrease together in the same proportion. In mathematics, this is expressed in a direct proportion. For example, if two employees assemble bicycles at the same speed, the number of bicycles they assemble is directly proportional to the time.

Inverse Proportionality: When two quantities are inversely proportional, it means that when one quantity increases, the other decreases in an inverse relationship. In mathematics, this is expressed in an inverse proportion. For example, if a car travels at a constant speed, the time it takes to reach its destination is inversely proportional to its speed.

Ratios and Rates: Proportions are used to compare two quantities, such as 2:3, which indicates that the first quantity is two-thirds of the second. Rates are common examples of proportions, such as interest rates or the flow rate of a river.

Percentages: Percentages are proportions expressed as a value from 0 to 100. For example, if 20% of the students in a school are members of a club, this is expressed as a ratio of 20:100 or 1:5.

Algebra and Geometry: Proportions are used in algebra and geometry to solve equations and problems related to length, area, and volume. They are also applied in trigonometry to express trigonometric relationships.

Problem Solving: Proportions are a valuable tool for solving mathematical and real-world problems that involve comparing quantities and proportional relationships. They can be used to calculate discounts, growth rates, mixing ratios, and much more.

Proportions are an integral part of mathematics and are applied in various situations to describe and solve problems involving proportional relationships between quantities or values.

Science: In scientific disciplines such as physics, chemistry, and biology, proportions are used to express relationships between physical quantities, such as speed, mass, concentration, and more.

Proportions play a fundamental role in science by allowing for the expression and understanding of relationships between different physical quantities.

Physics: Proportions are essential in physics for expressing relationships of speed, acceleration, force, and other quantities. For example, Newton's second law states that force is directly proportional to the acceleration of an object and its mass, which can be expressed through a proportion.

Chemistry: In chemistry, proportions are used to describe relationships in chemical reactions and to express concentrations of substances in solutions. The law of definite proportions states that chemical elements combine in fixed and proportional relationships.

Biology: In biology, proportions are useful for describing relationships between various quantities, such as the relationship between length and weight in animals, the concentration of substances in biological fluids, or growth relationships in populations.

Proportions are a universal tool for describing and communicating relationships between quantities in science and are essential for the development and understanding of scientific theories and experiments.

Economics: In economics, proportions are applied to analyze financial indicators, such as price-value relationships, interest rates, and economic growth.

Proportions are widely used to analyze and understand financial and economic indicators.

Price-Value Relationships: In the stock market, proportions such as the price-to-book (P/B) ratio, the price-to-earnings (P/E) ratio, and others are used to assess whether a company's shares are overvalued or undervalued relative to its financial fundamentals.

Interest Rates: Interest rates are expressed in terms of proportions, such as the annual interest rate expressed as a percentage of the principal amount borrowed. These rates are fundamental in the financial and economic realm for calculating financing costs and evaluating investments.

Economic Growth: Economic growth rates are expressed as proportions that compare the growth of a country's Gross Domestic Product (GDP) over different periods. These rates are essential for assessing a country's economic health and making economic policy decisions.

Price Indices: In inflation analysis, proportions are used to compare changes in price indices over different periods, allowing for the assessment of the impact of inflation on the prices of goods and services.

The use of proportions in economics is essential for making informed financial decisions, evaluating investment performance, and understanding economic trends in markets and economies at both national and international levels.

Statistics: Proportions are used in statistics to calculate and express relationships in data, such as the proportions of events in a dataset.

In statistics, proportions are a fundamental tool for analyzing and summarizing data.

Proportions and Probabilities: Proportions are used to calculate probabilities in random events. For example, in health statistics, proportions can be calculated to determine the likelihood of an event, such as a disease, occurring in a specific population.

Proportions and Rates: Rates and proportions are used to express relationships between subgroups of a population. For example, the mortality rate is a proportion that compares the number of deaths to the total population over a specific time period.

Group Comparisons: Proportions are used to compare the incidence of events across different groups. This is common in research studies, where treatment and control groups are compared to assess the effectiveness of an intervention.

Parameter Estimation: In statistical inference, proportions are used to estimate population parameters based on data samples. For example, the proportion of people in a sample with a specific characteristic is used to estimate the proportion in the entire population.

Proportions are a powerful tool in statistics for summarizing data, calculating probabilities, and comparing groups. They are used in a variety of fields, from health and economics to scientific and social research.

Geography: In cartography, proportions are used to represent the scale of a map and the relationship between distances on the map and in reality.

In geography and cartography, proportions, also known as "scales," are essential for representing and understanding the relationship between distances on a map and actual distances on Earth. Scales are commonly expressed in two ways:

Graphic or Linear Scale: This involves a line divided into distance units (e.g., kilometers or miles) on the map. By observing this line, you can determine how many distance units correspond to one unit of length on the map. For example, if a graphic scale shows that 1 centimeter on the map equals 10 kilometers in reality, you can easily calculate actual distances by measuring on the map.

Numerical Scale: This is represented as a ratio between one unit of distance on the map and one unit of distance in reality. For example, if the numerical scale is 1:100,000, it means that one unit of length on the map represents 100,000 times that length in reality. This scale is expressed as a fraction, where the first number represents the unit of length on the map, and the second number represents the unit of length in reality.

Scales are essential for the interpretation of maps, as they allow for distance estimation and understanding of the spatial relationship between places.

Additionally, they are crucial in urban planning, cartography, navigation, and other fields related to geography and representation of the Earth.

Education: Proportions are applied in the design of assessments and standardized tests, as well as in the comparison of grades and academic performance.

In the field of education, proportions play an important role in several areas:

Assessments and Standardized Tests: Proportions are used in the design of assessments and standardized tests to establish grading criteria. For example, a proportion may define how many questions must be answered correctly to achieve a specific grade. This ensures that tests are fair and consistent in evaluating student performance.

Grade Comparisons: Proportions are also applied in grade comparisons. For example, if you want to assess a student's performance relative to a grading scale, a proportion can be used to determine which grade range the student falls into. This is useful for evaluating academic performance and student progression.

Educational Data Analysis: In collecting and analyzing educational data, proportions can be used to examine relationships between variables, such as the ratio of male to female students in an educational institution, the proportion of students in different grades, and more. This provides valuable information for decision-making in school management.

Educational Program Design: When designing educational programs, proportions can be used to determine the ratio of students to teachers, the proportion of resources allocated to different curricular areas, and the distribution of learning opportunities.

In summary, proportions are a useful tool in the field of education for measuring performance, comparing outcomes, analyzing data, and making informed decisions about educational management and program design.

Health: In medicine and public health, proportions are used to describe the prevalence of diseases, the effectiveness of treatments, and other health indicators.

In the field of health and medicine, proportions are a fundamental tool for describing, analyzing, and communicating a variety of aspects related to public health, epidemiology, and clinical practice.

Disease Prevalence: Proportions are used to express the prevalence of diseases in a population, allowing for the determination of how many people are affected by a disease in relation to the total population. For example, the proportion of people with diabetes in a specific population.

Mortality and Survival Rates: Proportions are used to calculate mortality rates and survival rates in epidemiological studies. For example, the proportion of patients who survive after a specific treatment.

Evaluation of Treatment Effectiveness: Proportions are applied to assess the effectiveness of medical treatments and therapies. For example, the proportion of patients who respond positively to a new drug.

Infectious Disease Studies: Proportions are essential in the epidemiology of infectious diseases, as they are used to describe the spread and transmission of diseases, including infection rates and proportions of immunized individuals.

Clinical Research: In clinical research, proportions are used to measure the proportion of patients experiencing a side effect from a treatment or to assess the relationship between risk factors and clinical outcomes.

Quality of Medical Care: Proportions are applied in evaluating the quality of medical care, such as the proportion of patients receiving recommended treatments or adherence to care standards.

Public Health Resource Management: Proportions are useful in the allocation of resources in public health and in planning healthcare services based on the disease burden in a population.

Health Education: Proportions are used to communicate information about risks, benefits, and health outcomes to patients and the general public.

In summary, proportions are fundamental in health and medicine for quantifying and evaluating a wide range of aspects related to population health, treatment effectiveness, and the quality of medical care. These measures are essential for informed decision-making and the promotion of public health.

Engineering: In engineering, proportions are essential for the design and analysis of systems and structures, as well as for solving engineering problems.

In the field of engineering, proportions play a crucial role in various disciplines and applications.

Structural Design: In civil engineering and architecture, proportions are critical in the design of structures, such as bridges and buildings. Load and resistance proportions are key considerations to ensure that structures are safe and meet building codes.

Materials Mechanics: In this discipline, proportions are used to describe the relationships between stress and strain in materials such as steel and concrete. This is fundamental for the design of structural components and machines.

Thermodynamics: In mechanical and chemical engineering, proportions are applied to describe cycles of thermodynamic processes, such as refrigeration cycles and internal combustion engines.

Electronics and Circuits: In electrical engineering, proportions are used to define relationships of voltage and current in electrical and electronic circuits. This is essential for the design and analysis of electronic systems.

Control Engineering: In process automation and control, proportions are applied to describe the relationships between input and output signals in control systems. This is critical in the regulation of industrial systems and robotics.

Software Engineering: In software engineering, proportions are used to define metrics of software quality and performance, such as the proportion of defective lines of code relative to the total lines of code.

Transportation Engineering: In the planning of transportation systems and road design, proportions are applied to describe traffic flow rates and the capacity of transportation infrastructures.

Environmental Engineering: Proportions are used in the management of natural resources and environmental impact assessment to describe relationships between the quantity of available resources and the demand for natural resources.

Aerospace Engineering: In the design of aircraft and rockets, proportions are critical for determining the relationship between speed, efficiency, and payload capacity of spacecraft.

Systems Engineering: In systems engineering, proportions are applied to define relationships between components and subsystems in complex systems.

In each of these engineering disciplines, proportions help quantify and understand the relationships between various variables and are essential for design, analysis, and problem-solving in engineering.

Business: In business management, proportions are applied in financial analysis, decision-making, and evaluating business performance.

In the world of business and corporate management, proportions play a crucial role in various aspects.

Financial Analysis: Financial ratios, such as the debt-to-equity ratio, profit margin, liquidity ratio, and inventory turnover, are used to assess a company's financial health. These ratios enable investors, shareholders, and financial managers to make informed decisions about resource management.

Planning and Budgeting: Proportions are used to set financial goals and create business budgets. This involves determining the relationships between income and expenses, which is essential for maintaining financial balance.

Performance Evaluation: Proportions are applied to assess a company's performance in comparison to its competitors or its own past results. This can include metrics such as profitability, operational efficiency, and growth.

Company Valuation: Proportions are used in company valuation to determine intrinsic value. This is relevant in mergers and acquisitions, as well as in the sale of businesses.

Inventory Management: Ratios, such as inventory turnover, help companies manage their inventory levels efficiently. This ensures there is enough inventory to meet demand without incurring excessive costs.

Human Resources: Proportions are applied in human resource management to evaluate workforce efficiency. This may include the ratio of employees per supervisor, staff turnover, and other metrics related to personnel performance.

Marketing and Sales: Proportions are used in marketing and sales to evaluate the return on investment (ROI) of marketing strategies and the effectiveness of sales efforts. This helps determine which strategies are most profitable.

Project Management: Proportions are applied in project management to assess progress and project performance against established goals and budgets.

Sustainability: Proportions related to sustainability, such as the ratio of renewable energy used compared to total energy consumed, are fundamental for corporate social responsibility initiatives and environmental management.

Strategic Decision-Making: Financial and operational proportions are used to make strategic decisions, such as company expansion, product diversification, or entry into new markets.

In summary, proportions are essential tools in business management for measuring performance, evaluating efficiency, and making informed decisions that affect the health and growth of a company. These metrics provide a quantitative view of various aspects of business management and are fundamental for financial and strategic analysis.

19.Arithmetic Mean: The average of a set of numbers

The arithmetic mean, commonly known as the average, is a statistical measure used to represent the typical or central value of a set of numbers. To calculate the arithmetic mean, all the numbers in the set are summed and then divided by the number of numbers in that set. The formula for the arithmetic mean is expressed as follows:

Arithmetic Mean = (Sum of the numbers) / (Number of numbers)

For example, let's consider the set of numbers: 5, 7, 9, 12, 15. To calculate the arithmetic mean of this set, we first sum the numbers:

5 + 7 + 9 + 12 + 15 = 48

Then we divide the sum by the number of numbers in the set, which is 5 in this case:

Arithmetic Mean = 48 / 5 = 9.6

Therefore, the arithmetic mean of this set of numbers is 9.6. This means that 9.6 is the average value of the numbers in the set. The arithmetic mean is a commonly used measure in statistics and mathematics to summarize data and understand trends or central values in a set of numbers.

Identifying Trends: The arithmetic mean allows analysts to identify trends in a data set. By calculating the average, one obtains a value that represents the "center" of the data. This means that if the mean is significantly higher or lower than certain values in the data set, one can conclude that there is a trend in that direction. For example, in a monthly sales data set, a mean that increases over time indicates growth in sales.

Detecting Outliers: Outliers or extreme values in a data set can influence conclusions and decisions. The mean can help detect these outliers. If there is an extreme value in the data set, the mean can be significantly affected. This alerts analysts to the presence of unusual values. For example, in a data set of employee salaries, an extremely high or low salary would cause the mean to differ from what is considered typical.

Detecting outliers or extreme values in a data set is one of the most important applications of the arithmetic mean in data analysis. Here is more information on this aspect:

Identification of Outliers: When calculating the arithmetic mean, a value is obtained that represents the center of the data set. Values that deviate significantly from this mean can be considered outliers or extremes. These values may be higher or lower than expected based on the typical behavior of the data.

Importance of Outlier Detection: Detecting outliers is crucial in many contexts, as these values can distort conclusions and data-based decisions. In areas such as medical research, financial risk assessment, or product quality, the presence of outliers can indicate significant problems or opportunities.

Examples of Outlier Detection: Imagine a data set of student grades on an exam. If most students score in the range of 70-90, but one student scores 10, that low score is an outlier. In a stock investment context, if most stocks in a portfolio increase in value but one stock decreases significantly, that stock might be considered an outlier.

Actions Following Outlier Detection: Once outliers are identified, it is important to decide what to do with them. In some cases, outliers may be data entry errors and should be corrected. In other cases, they may indicate significant events that require further investigation. Depending on the context, they may be removed, adjusted, or analyzed in detail.

In summary, the arithmetic mean is a valuable tool for detecting outliers in a data set, helping to ensure that data-based conclusions and decisions are more accurate and reliable.

Comparing Results: The arithmetic mean is useful for comparing results between different data sets or groups. For instance, if a company wants to compare the performance of two sales teams in different regions, calculating the average sales of each team allows for a direct comparison. The team with the highest average sales would be considered more effective.

Indeed, the arithmetic mean is a fundamental tool for comparing results and evaluating performance in various contexts. Here is more detail on its usefulness in comparing results:

Comparison of Groups or Data Sets: The arithmetic mean provides a representative value for each group or data set. This allows for simple and direct comparisons between groups. For example, in the case of sales teams in different regions, the means of the sales of each team can be calculated and then compared to evaluate which team has a better average performance.

Assessment of Trends Over Time: In addition to comparing groups, the arithmetic mean is also used to assess performance over time. For example, if a company wants to analyze whether sales have increased or decreased over the past five years, it can calculate the average sales for each year and observe if there is a general upward or downward trend.

Informed Decision-Making: By comparing results through the arithmetic mean, organizations and decision-makers can base their actions on objective data and quantitative analysis. This can be crucial in strategic planning, resource allocation, and operational decision-making.

Evaluation of Deviations or Discrepancies: Comparisons through the arithmetic mean can also reveal significant deviations. For example, if one sales team has a much higher average performance than others, this discrepancy may indicate the need to investigate what practices or strategies are driving that exceptional performance.

In summary, the arithmetic mean is a versatile tool that facilitates the comparison of results in a variety of situations, enabling organizations and decision-makers to take informed and strategic actions based on quantitative data.

Informed Decision-Making: In a variety of fields, from economics to medicine, decision-making is based on data. The arithmetic mean provides a central reference point that aids in making informed decisions. For example, in healthcare, data analysis may involve comparing patient outcomes in clinical trials. The average of the results can be crucial in determining whether a treatment is effective.

Informed decision-making is essential across a wide range of fields, and the arithmetic mean plays a key role by providing a central reference for evaluating and understanding data. Below is a detailed explanation of how the arithmetic mean facilitates informed decision-making in various contexts:

Economics and Finance: In investment and economic contexts, the arithmetic mean is used to calculate indices and growth rates. For example, calculating the average return of an investment portfolio over time provides critical information for investors. Informed decisions regarding where to allocate financial resources are largely based on these performance measures.

Medicine and Public Health: In medical research, clinical trials, and epidemiological studies, the arithmetic mean is used to summarize data on treatment efficacy, disease prevalence, and other health indicators. Doctors and researchers use these statistics to make informed decisions about treatments and public health policies.

Education: In the educational field, the arithmetic mean is used to evaluate student and school performance. Educators and decision-makers use this data to identify areas for improvement and to design effective educational strategies.

Resource Planning: In both business management and natural resource planning, the arithmetic mean is used to allocate resources efficiently. For example, in business logistics, the average product demand can be used to plan inventory and supply chains.

Politics and Government: In policy formulation and government decision-making, data collected and summarized through the arithmetic mean can support the implementation of public policies. For example, by calculating the average family income, governments can make informed decisions regarding fiscal policies and social assistance programs.

In all these cases and many others, the arithmetic mean serves as an essential benchmark that enables professionals and decision-makers to understand and analyze data, which in turn supports informed and strategic decision-making across a wide variety of fields.

In summary, the arithmetic mean is a tool for summarizing data and better understanding its nature. It helps analysts and decision-makers draw conclusions, identify patterns, and compare results, which is essential for data-driven decision-making across a wide range of fields.

20.Medians: The middle number in an ordered set of numbers

The median is a statistical value used to represent the central number in an ordered data set. To calculate the median, you must first sort the data in ascending or descending order. Then, if you have an odd number of data points, the median is the value that occupies the central position in the ordered sequence. If you have an even number of data points, the median is the average of the two central values.

How to Calculate the Median:

Sort the data: The first step to calculating the median is to sort the data set in ascending or descending order, which makes it easier to identify the central value. For example, if you have the following data set: 7, 2, 1, 9, 5, you should sort them in ascending order: 1, 2, 5, 7, 9.

Odd number of data points: If the data set has an odd number of elements (for example, 5), the median is simply the value that occupies the central position in the ordered sequence. In this case, the third value (5) is the median, as it divides the data into two equal halves, with two smaller values and two larger values.

Even number of data points: If the data set has an even number of elements (for example, 6), the median is calculated by averaging the two central values. In this case, the two central values are the third and fourth values after sorting. You must add these two values and then divide the sum by 2 to get the median.

For example, if you have the data set: 2, 4, 6, 8, the median would be calculated as follows:

Sort the data: 2, 4, 6, 8.

Since you have an even number of data points, take the two central values, which are 4 and 6.

Average of the central values: (4 + 6) / 2 = 5.

So, the median of this data set is 5.

The median is especially useful when working with data that do not follow a normal distribution and can be less sensitive to outliers compared to the arithmetic mean.

Because the median is based on the central position of ordered data, it is less sensitive to extreme or outlying values in the data set compared to the arithmetic mean. This makes it a robust measure of central tendency in situations where extreme values can skew the mean.

The median is particularly useful in the following cases:

When working with skewed or asymmetric data sets.

When an asymmetric data set or a skewed distribution is mentioned, it refers to situations where the frequency of values is not evenly distributed along a scale. In other words, there is a concentration of data around certain values or ranges, while other values may be less common and, at times, extremely unusual. This creates asymmetry in the data distribution.

In such cases, the median is useful because it is not significantly affected by extreme or outlier values, which often lie in the tail of the distribution. Since the median is based on the central position of ordered data, it focuses on the values that occupy a middle position rather than considering the average of all values, which could be skewed by extreme values.

For example, in a distribution skewed to the right (where most values are on the lower end), the mean could be inflated by a few extremely high values. In this case, the median would be a more representative indicator of "central tendency" because it is based on the central values in the ordered sequence, making it less susceptible to the influence of outliers.

In summary, in situations with skewed or asymmetric distributions, the median is considered a more robust and useful measure of central tendency than the arithmetic mean.

When outliers can significantly impact the mean and a more resistant measure to these extreme values is desired.

The median is a measure resistant to outliers, which makes it especially valuable in situations where extreme values can significantly affect the arithmetic mean and distort data interpretation.

Outliers or extreme values are observations that are much higher or lower than most other values in a data set. In the case of the arithmetic mean, these values can have a disproportionately large impact, since they are averaged with all other values. This can distort the central tendency and lead to incorrect conclusions about the nature of the data.

The median addresses this issue by focusing on the value that occupies a central position when the data is sorted. Since the median only takes into account the central value or the two central values (in the case of an even number of data points), it is not significantly influenced by outliers. Instead, it better reflects the "central tendency" of the data, especially when the data contains extreme values that are not representative of the majority.

In summary, if a measure of central tendency is needed that is resistant to outliers and provides a more accurate representation of the data's "trend" in the presence of extreme values, the median is a suitable choice.

In situations where only ordinal or interval data are available, and mathematical operations on the actual values cannot be performed.

The median is especially useful in situations where data are presented on an ordinal or interval scale, and where mathematical operations such as addition and subtraction do not make sense or are not appropriate. Since the median is based solely on the relative position of values within the data set, it does not require any numerical computation of the values themselves. This makes it applicable in a wide range of contexts where the values are only comparable but not quantifiable in absolute terms.

For example, consider a survey where participants are asked to rate their satisfaction with a product on an ordinal scale, with options such as "very dissatisfied," "dissatisfied," "neutral," "satisfied," and "very satisfied." In this case, you cannot perform mathematical operations on the ratings, since they do not represent exact numerical quantities. However, you can calculate the median of the ratings to obtain a measure of the central rating that reflects how most respondents felt about the product.

In summary, the median is a valuable tool when working with ordinal or interval data and mathematical operations on the underlying values cannot be performed. It provides a measure of central tendency based solely on the relative position of the data, making it applicable in a variety of research and data analysis contexts.

Its applicability across a wide range of fields—from economics to medicine and beyond—makes it an essential tool for summarizing and understanding information contained in diverse data sets.

The median is a versatile and widely used measure of central tendency that applies to numerous fields due to its ability to effectively summarize and interpret data. Its resistance to outliers, its applicability to skewed or asymmetric data sets, and its ability to work with ordinal and interval data make it an essential tool in research, data analysis, and decision-making across a variety of disciplines.

Whether in economics, medicine, education, social research, or any other field, the median provides valuable insight into the central tendency of data, facilitating pattern recognition and informed decision-making.

The median is a fundamental tool in a wide range of fields, as it provides a measure of central tendency that supports pattern recognition and informed decision-making based on data. Being less sensitive to outliers compared to the arithmetic mean, the median becomes a valuable option when working with data sets that may contain extreme values. This makes it especially relevant in contexts where the robustness of the central tendency measure is crucial for obtaining an accurate assessment of a situation.

Whether in evaluating economic outcomes, measuring the effectiveness of medical treatments, or comparing performance in education, the median plays a key role in understanding and analyzing data.

The median is an essential tool in data evaluation and analysis across various fields, including economics, medicine, and education. Its ability to provide a measure of central tendency that is not significantly affected by outliers makes it a valuable choice when working with data that may contain variability or extremes. In economics, it is used to understand income distribution and assess inequalities. In medicine, it is applied to measure treatment effectiveness and evaluate health outcomes. In education, it is used to compare student or school performance. In all these contexts, the median provides important information that contributes to more well-founded decision-making.

21.Mode: The number that appears most frequently in a set of numbers

Mode in Statistics and Its Applications

In statistics, the mode refers to the number or value that occurs most frequently in a data set. There can be a single mode (a value that appears most often) or multiple modes if several values have the same highest frequency. The mode is a measure of central tendency used to summarize data and describe the distribution of values in a data set.

To calculate the mode, one must identify the value or values that appear most frequently in the dataset. The mode is especially useful when working with categorical or discrete data, such as product categories, integers, or grade categories. It is applied in various fields, such as statistics, sociology, epidemiology, fashion, and market research to identify patterns and trends in data.

The mode is a particularly useful statistical measure when dealing with data that is categorical or discrete — that is, data divided into categories or having a finite set of possible values.

Product Categories:

In the field of fashion and retail, the mode is used to identify the most popular products in a catalog. For example, determining which clothing color sells most frequently in a store. In fashion and retail, the mode plays a crucial role in identifying consumer trends and preferences.

Inventory Selection:

Retail stores use the mode to determine which products to keep in stock and which might require promotions or be removed from inventory. For example, if blue jeans are found to be the most popular mode in a clothing store, the store may choose to increase its stock of blue jeans.

Seasonal Planning:

The mode is also essential for seasonal planning. Retailers must anticipate which colors, styles, or products will be popular in a specific season, such as summer or winter, to meet consumer demand.

Product Design:

Fashion designers and manufacturers use mode data to create products that align with consumer preferences. For example, if a certain pattern or print is the most popular mode, fashion companies may incorporate that design into their product lines.

Marketing and Advertising:

Mode results are often used in marketing and advertising campaigns. If a color or style is very popular, brands will highlight it in their ads to attract buyers.

Trend Prediction:

Mode helps retailers and designers forecast future trends in the fashion industry. This information is valuable for making decisions about which products to develop and promote in the future.

In the world of fashion and retail, the mode is a tool that supports decision-making related to product development, inventory, and marketing. It enables businesses to satisfy consumer preferences and remain competitive in the market.

Grades and Academic Performance:

In education, the mode can be used to identify the grade most commonly achieved by students on an exam or in a class, providing insights into overall performance.

Performance Assessment:

The mode of grades can reveal the most common level of achievement among students in a specific evaluation. For example, if a grade of "B" is the mode on an exam, it indicates that most students received a "B" grade.

Identifying Strengths and Weaknesses:

Teachers and educators can use the mode of grades to identify areas where students excel and areas where they may need more support. If a specific grade is the mode, it may indicate a common strength or weakness among students.

Course Content Evaluation:

The mode of grades can also help educators evaluate the effectiveness of their teaching and course content. If a particular grade is the mode but does not reflect the intended learning outcomes, the curriculum or teaching methodology may need revision.

Long-term Trend Detection:

By observing grade modes over time, educational institutions can identify trends in student performance. This can be useful for curriculum planning and spotting areas for improvement.

Motivation and Feedback:

Showing students the mode of grades can serve as feedback on their performance. Students can use this information to motivate themselves to improve and to understand how they compare to their peers.

The mode of grades is a tool that helps educators and students better understand academic performance and make informed educational decisions.

Sociology:

In sociological studies, the mode is used to analyze categorical data such as political preferences, religion, occupation, or affiliations. For example, it can help identify the most common religion among the inhabitants of a region.

In sociology, the mode is used to analyze categorical data related to society and culture.

Identifying Cultural Trends:

The mode is used to identify cultural trends in a population or region. For example, determining the predominant religion in a community or geographic region. This helps sociologists understand religious diversity in society.

Studying Political Preferences:

Sociological studies can use the mode to analyze the political preferences of a population. For instance, identifying the most popular political party among voters in a region.

Occupation Analysis:

The mode is also applied to analyze the most common occupations or professions in a society. This helps to understand the labor and economic structure of a region.

Affiliations and Memberships:

In sociology, the mode is used to study group or association affiliations. For example, determining how many citizens in a community enjoy participating in volunteer organizations.

Changes Over Time:

The mode can be analyzed over different time periods to identify changes in preferences, beliefs, or affiliations in a society. This helps to understand the evolution of culture and society.

In general, the mode is a useful tool for sociologists seeking to understand diversity and cultural trends across different communities and populations. It helps identify patterns in categorical data relevant to sociology and social research.

Epidemiology:

In public health and epidemiology, the mode can be used to identify the most frequent symptoms or diseases in a population group, which is crucial for disease detection and control.

The mode is a useful tool in epidemiology and public health for identifying the most common symptoms or illnesses in a population. This is important for several purposes:

Outbreak Detection:

The mode can help quickly identify the most frequent symptoms in a population, which is crucial for early detection of disease outbreaks. If a symptom or illness becomes notably more common than expected, it may signal a need to investigate and control the spread of the disease.

Health Resource Planning:

Identifying the most common illnesses or symptoms allows health authorities to plan for medical resources and services. For example, if certain symptoms are prevalent in a region, appropriate resources for diagnosis and treatment can be allocated.

Monitoring Health Trends:

The mode can be used to track health trends in a population over time. This is useful for monitoring changes in the prevalence of certain diseases or symptoms and assessing the effectiveness of public health interventions.

Risk Factor Identification:

Identifying the most common symptoms or diseases can also help determine risk factors and underlying causes of health problems. This is fundamental to epidemiological research.

Evaluation of Intervention Effectiveness:

The mode is used to assess whether public health interventions are impacting the reduction of specific symptoms or diseases in a population.

The mode helps healthcare professionals detect emerging diseases, plan medical care, and evaluate the impact of public health interventions.

Market Research:

In market research and consumer analysis, the mode is used to identify consumer preferences. For example, it can help determine the most popular ice cream flavor in a region. In market research and consumer analysis, the mode helps identify consumer preferences and make strategic decisions. By determining the product or product variant that is most popular or frequently chosen by consumers, companies can:

Develop Marketing Strategies:

Knowing the mode allows companies to tailor their marketing and advertising strategies to highlight products or features that are most popular among consumers.

Inventory Planning:

Companies can plan their inventory based on the mode, ensuring they have enough stock of the most in-demand products.

Product Development:

The mode can also guide the development of new products. If a specific flavor or variant is especially popular, a company might choose to develop more products in that category or expand its product line in that direction.

Production Decisions:

Knowing the mode can help companies decide how much of a product to produce. For example, if a certain ice cream flavor is very popular, the company might increase its production.

Competition and Differentiation:

Understanding the mode also allows companies to compete more effectively. They can differentiate themselves from competitors by offering products or features that are particularly popular among consumers.

The mode is an important tool in market research and consumer analysis, enabling companies to make informed decisions about products, marketing strategies, and inventory planning based on consumer preferences.

22.Exponents: Small numbers written above and to the right of other numbers, like 2^3, which means 2 x 2 x 2

Exponents, also known as powers, are a mathematical notation used to indicate the number of times a number, called the "base," is multiplied by itself. In the expression "2^3," the base is 2, and the exponent is 3. This means that 2 is multiplied by itself three times, which can be expressed as: 2^3 = 2 × 2 × 2 = 8 Thus, 2^3 equals 8. Exponents are fundamental in mathematics and are used in a wide variety of concepts, from basic arithmetic to more advanced calculations in algebra, calculus, and number theory. Exponents have many applications in areas such as physics, engineering, computer science, and science in general, where they are used to describe and understand phenomena involving exponential growth, rates of change, and more.

Exponents play a crucial role in various scientific disciplines and practical applications. Here is a more detailed explanation of their importance across several fields:

Physics: In physics, exponents are used to describe phenomena of exponential growth, such as radioactive decay, the decay of subatomic particles, and population growth. Additionally, the laws of thermodynamics and the theory of relativity include equations with exponents that model fundamental relationships in the universe. Exponents are essential for describing a variety of phenomena involving exponential growth, rates of change, and decay.

Radioactive Decay: The radioactive decay of atomic nuclei follows exponential kinetics. The number of remaining radioactive atoms in a sample decreases over time at a constant rate. Exponents are used to describe the decay rate and calculate the half-life of radioactive nuclei.

Population Growth: In particle physics and astrophysics, exponents are employed to model the growth and decay of subatomic particles, as well as to understand how the particle population density in the universe changes.

Exponential Growth in Electronics: In electronics, the charging and discharging of capacitors and the charging and discharging of RC circuits follow exponential curves. Exponents are used to describe how currents and voltages change over time.

Thermodynamics: Thermodynamics, a fundamental branch of physics, uses equations with exponents to model the relationships between temperature, pressure, and volume of ideal gases, as well as to predict changes in internal energy and entropy of thermodynamic systems.

Oscillatory and Wave Phenomena: In wave and oscillatory phenomena, such as the oscillations of a simple pendulum, equations may involve exponents, especially in solutions to differential equations, to describe motion over time.

Theory of Relativity: Einstein's theory of relativity includes equations involving exponents. For example, the equation that describes time dilation due to velocity is expressed with an exponent that depends on relative speed.

Heat Transfer Phenomena: The diffusion of heat through materials is governed by differential equations involving exponents. These equations describe how temperature changes with time and position.

Stellar Nuclei and Supernovae: The physics of stars and nuclear processes in stellar cores and supernovae involves equations with exponents to model the decay and fusion of nuclear elements.

In all these contexts, exponents are used to understand and predict how certain physical quantities change with respect to time, position, and other variables. They are fundamental for modeling natural phenomena and for problem-solving in theoretical and applied physics.

Engineering: Exponents are vital in engineering to describe the behavior of physical and electrical systems, such as electrical circuits, control systems, and heat diffusion in materials. They are also applied in engineering to model the growth of structures, wave propagation, and fluid flow.

Exponents play a crucial role in engineering, as they are used to describe and predict the behavior of physical and electrical systems, as well as to model a variety of phenomena and processes.

Electrical Circuits: In electrical and electronic engineering, exponents are applied in the analysis of electrical circuits, especially in capacitor charging and discharging systems and in alternating current circuits. Exponents help describe how currents and voltages change over time.

Control Systems: In control engineering, exponents are used in the modeling and analysis of control systems, such as feedback systems and automatic control. They help predict how state variables evolve and how a system responds to changes in inputs.

Heat Transfer: Mechanical engineering and materials engineering apply exponents in heat transfer and temperature diffusion through materials. Heat diffusion equations involve exponents to describe how heat propagates in solids, liquids, and gases.

Structures and Materials: Civil and structural engineers use exponents in modeling the growth of structures and in analyzing stresses and strains. They are also applied to predict the lifespan of materials subjected to cyclic loads.

Wave Propagation: In telecommunications and antenna design, exponents are employed to model the propagation of electromagnetic waves. In acoustics, they are used to describe the propagation of sound waves in various media.

Fluid Dynamics: Mechanical and aerospace engineering apply exponents in fluid dynamics to describe the flow of liquids and gases in conduits and through structures. This is crucial for the design of piping systems, aircraft, rockets, and spacecraft.

Structural and Geotechnical Engineering: Civil engineering involves the use of exponents in the analysis of foundations, soil consolidation, and seismic risk assessment. These exponents help understand how loads and stresses propagate in structures and soils.

Systems Engineering: Systems engineers apply exponents in the analysis of complex systems, such as communication networks and transportation systems. They help predict the performance of systems over time and interactions.

In summary, exponents are an essential mathematical tool in engineering that allows engineers to model, analyze, and understand a wide variety of phenomena and systems in fields as diverse as electronics, mechanics, energy, communication, and construction. These applications are fundamental for the design, analysis, and optimization of systems and devices in engineering.

Computer Science: In computer science and computing, exponents are fundamental for analyzing algorithmic complexity. Efficient algorithms are often expressed in terms of exponents, allowing for the evaluation of their effectiveness in terms of time and resources.

In computer science and computing, the use of exponents plays a critical role in evaluating and analyzing algorithmic complexity.

Exponent Notation in Algorithmic Complexity: Exponent notation is a way of expressing the time and space complexity of algorithms. For example, it is common to describe the time complexity of an algorithm using "$O(n^x)$" notation where "n" is the size of the dataset and "x" is an exponent that indicates how execution time grows with the size of the data. Common examples include "O(n)" (linear), "O(n log n)" (linearithmic), "O(n^2)" (quadratic), and "O(2^n)" (exponential).

Algorithm Efficiency Analysis: Exponents are used to measure the efficiency of algorithms. The smaller the exponent value, the more efficient the algorithm. This is critical for designing algorithms that solve problems quickly and efficiently, which is essential in computing.

Space Complexity: In addition to time complexity, exponents are also used to describe the spatial complexity of algorithms. This refers to the amount of memory or resources required to execute an algorithm based on the size of the input data. Analyzing spatial complexity helps optimize resource use in computing.

Complexity Classes: Complexity classes such as P, NP, NP-complete, and NP-hard are defined in terms of exponents. These complexity classes are fundamental in computational theory and are used to classify problems based on their computational difficulty.

Algorithm Optimization: Exponents are also applied in optimizing algorithms. Algorithms that can reduce a problem's complexity to a lower exponent are considered more efficient and valuable.

Resource and Time Planning: In software development and project planning in computing, analyzing algorithmic complexity, which involves exponents, is fundamental for estimating the time and resources needed to complete a task or project.

Algorithm Comparison: Exponents allow for comparing the efficiency of different algorithms for a given task. This is essential when selecting the most suitable algorithm for a specific application.

In summary, exponents are an essential tool in computer science for analyzing, comparing, and optimizing algorithms. They provide a quantitative way to evaluate the efficiency of algorithms in terms of time and resources, which is crucial in a field where efficiency is vital for solving problems quickly and effectively.

Biology: In biology, exponents are used to describe population growth, DNA replication, and the spread of infectious diseases. Exponential models are common in biology to understand how populations of organisms change over time.

In biology, exponents play a crucial role in describing and modeling a wide variety of phenomena and processes.

Population Growth: Exponents are used to describe the growth of populations of organisms. A common model is exponential growth, expressed as "N(t) = N0 * e^(rt)," where "N(t)" is the population at time "t," "N0" is the initial population, "e" is the base of the natural logarithm, "r" is the growth rate, and "t" is time. This model is fundamental for understanding how biological populations change over time.

DNA Replication: In molecular biology, exponents are used to describe DNA replication. Replication is an exponential process in which a DNA strand duplicates to form two identical strands. Exponential notation is used to represent the exponential amplification of genetic material.

Spread of Infectious Diseases: In epidemiology, exponential models are used to describe the spread of infectious diseases. These models consider factors such as the infection rate, incubation period, and basic reproduction number (R0) to predict how a disease will spread through a population.

Bacterial Growth: Exponents are also applied to describe the growth of bacteria and other microorganisms. Bacterial growth follows an exponential model under ideal conditions, which is fundamental for microbiology and biotechnology.

Genetic Mutations: Genetic mutations can be modeled with exponents. Mutation rates are expressed as the probability of a mutation occurring in a gene or genetic locus over a given time, which can be represented using exponential notation.

Population Evolution: Changes in allele frequencies in a population over time are often described using models based on exponents, such as the Hardy-Weinberg model. This model helps understand how allele frequencies change in a population at equilibrium.

Ecology: Exponents are applied in ecology to model species growth, population dynamics, and interactions among organisms in an ecosystem. Exponential and logistic models are fundamental to this field.

Toxicology: In toxicology, exponents are used to describe how the concentration of a toxic substance in an organism changes over time. This is relevant for understanding the kinetics of toxin elimination from the body.

In summary, exponents are an essential tool in biology for modeling a variety of processes and phenomena, from population growth to DNA replication and the spread of diseases. These mathematical models are fundamental for understanding and predicting biological events and supporting research and applications in biology-related fields.

Economics: Economics uses exponents to model economic growth, inflation, interest rates, and investment. Concepts of compound interest and net present value in finance involve exponential calculations.

In the field of economics and finance, exponents are fundamental for modeling a series of key concepts and processes. Here are some of the most common applications of exponents in economics:

Economic Growth: Economic growth models often rely on exponential functions or exponential growth, which describe how a country's output increases over time. These models are fundamental for analyzing economic development and long-term economic projections.

Inflation: Inflation, representing the widespread increase in prices of goods and services, can be modeled using an annual inflation rate. Exponents are used to calculate the future value of a sum of money after a period of inflation.

Interest Rates: Compound interest rates play a crucial role in finance and economics. Exponents are used to calculate the future value of an investment or loan when compound interest rates are applied. The formula for compound interest is $V = P(1 + r/n)^{(nt)}$, where "P" is the principal, "r" is the annual interest rate, "n" is the number of times interest is compounded per year, "t" is time, and "V" is the future value.

Net Present Value (NPV): In investment decision-making and project evaluation, NPV is a measure that assesses the profitability of an investment over time. This measure involves discounting future cash flows using discount rates, which are often expressed with exponents.

Asset Depreciation: In economics and accounting, fixed assets such as machinery or vehicles may depreciate over time. Exponents are used to calculate the annual depreciation of these assets.

Investment Growth: In investment and financial planning, exponents are used to assess the potential growth of investments over time. This is crucial for determining investment strategies and retirement planning.

Game Theory: In game theory, exponents are used to model competitive and strategic situations. Game models may involve exponential calculations to evaluate strategies and outcomes.

Business Valuation: The valuation of companies and assets, such as stocks and bonds, involves financial calculations with exponents. Financial analysts use valuation models to determine the intrinsic value of assets and companies.

Equilibrium in the Goods Market: In macroeconomics, exponents are used in models that describe equilibrium in the goods market. These models help understand how investment and savings can balance at a macroeconomic level.

Financial Decision-Making: In financial decision-making situations, such as investments, loans, and retirement planning, exponents are fundamental for calculating the future value or present value of future cash flows.

In summary, exponents play a critical role in economics and finance, as they are used to model economic growth, calculate interest rates, assess investments, and make informed financial decisions. These mathematical concepts are essential for analysis and decision-making in the economic and financial realm.

Environment and Earth Sciences: Geology and climatology often use exponents to analyze the growth and degradation of natural resources, climate change, and soil erosion.

The use of exponents in the field of environmental and Earth sciences is fundamental for analyzing various natural processes and phenomena related to geology and climatology.

Natural Population Growth: In ecology and biology, exponents are used to describe the growth of populations of organisms in ecosystems. Exponential growth models are employed to understand how populations of plants and animals increase over time, which is crucial for conservation and resource management.

Soil Erosion: Soil erosion is a process involving the gradual loss of the top layer of soil due to factors such as rain and wind. Scientists use exponential models to estimate the rate of soil erosion and its impact on agriculture and the environment.

Depletion of Natural Resources: In geology, exponents are applied to study the degradation of natural resources, such as the depletion of mineral reserves or coastal erosion. These models are fundamental for assessing the sustainability of resource extraction.

Climate Change: In climatology and environmental sciences, exponents are used to model climate change and the increase in greenhouse gas concentrations in the atmosphere. Exponential models help predict the impact of climate change in terms of temperature, sea level rise, and extreme weather events.

Ecosystem Recovery: When addressing restoration and recovery efforts for degraded ecosystems, exponents are used to predict the time required for an ecosystem to recover to a natural or healthy state.

Tectonic Plate Dynamics: In geology, exponents may describe the speed of tectonic plate movement and the generation of earthquakes and volcanic activity. These models are relevant for understanding Earth's dynamics and geological processes.

Glacier Retreat: Scientists study glacier retreat using exponential models to estimate the rate at which glaciers are shrinking due to climate change.

Coastal Erosion: In coastal areas, coastal erosion is a significant problem. Exponential models are used to predict the loss of coastal land and to develop coastal management strategies.

In summary, exponents are a valuable tool in research and analysis of natural processes and environmental phenomena. They help scientists better understand how the Earth and its environment are changing over time and to take action to preserve and protect our natural resources.

Statistics: In statistics, exponents apply to probability distributions, especially exponential and Poisson distributions. These statistical models are used to describe random events, such as the time between customer arrivals at a service.

Exponential Distributions: Exponential distributions are widely used in statistics to model the probability of an event occurring within a specific time period. Exponents are present in the probability density function of the exponential distribution. This distribution is used in problems such as the time between customer arrivals at a service (Poisson process) and the lifespan of electronic devices.

Poisson Distribution: The Poisson distribution is another statistical distribution related to random events. Exponents are used in the probability function of the Poisson distribution to model the occurrence of events in specific intervals of time or space. Examples of applications include the frequency of phone calls to a call center or the incidence rate of traffic accidents in a region.

Waiting Time and Queues: In queue theory problems, which apply to waiting systems such as customer service lines and service processes, exponents are used to model the time customers wait in line before being served. This is crucial for optimizing service processes and resource management.

Reliability and Maintenance: Exponents also apply in the analysis of system reliability and maintenance. They help model the time until a system fails or requires maintenance and to make informed decisions about when to carry out preventive maintenance tasks.

Survival Statistics: In survival analysis, exponential models are used to study the duration of life of objects, patients, or any entity that may experience an event of interest (such as product failure or patient death). Exponents are integral to proportional hazard models and are used to estimate the survival function.

In summary, exponents play a crucial role in analyzing random events and stochastic processes in statistics. These models are essential for describing and predicting the time between events and provide a solid foundation for decision-making in a variety of fields, from customer service to medical research and logistical planning.

Mathematics: In mathematics, exponents are an essential component of algebra and calculus. Exponential and logarithmic functions are fundamental in mathematical analysis and in solving differential equations.

Algebra: Exponents are used to express power operations. For example, in the expression "2^3," the exponent 3 indicates that 2 is multiplied by itself three times. This is fundamental in basic arithmetic and algebra.

Properties of Exponents: In algebra, the properties of exponents are studied, including the multiplication of powers of the same base (such as a^m * a^n = a^(m+n)) and the division of powers of the same base (such as a^m / a^n = a^(m-n)). These properties are essential in simplifying algebraic expressions and solving equations.

Exponential Functions: Exponential functions, such as f(x) = a^x, are a fundamental topic in algebra and calculus. These functions represent exponential growth and are used to model phenomena in natural sciences, economics, and more. The exponents in these functions determine how quantities change as x varies.

Logarithms: Logarithms are inverses of exponential functions and are crucial in algebra and calculus. They help solve exponential equations and simplify expressions. Exponents are used in the definition of logarithms as $\log_a(b) = x$, where "a" is the base, "b" is the value, and "x" is the exponent.

Calculus: Exponents appear in calculus in the context of derivatives and integrals. The derivatives of exponential and hyperbolic functions, as well as their properties, involve exponents. Exponents are also used when solving differential equations that model physical and natural phenomena.

Scientific Notation: Scientific notation, which involves the use of exponents, is fundamental in mathematics and sciences. It helps express very large or very small numbers in a more compact and manageable way.

In summary, exponents are an essential part of mathematics and are applied in a variety of contexts, from basic arithmetic operations to exponential and logarithmic functions in calculus. Their understanding and manipulation are fundamental for advancing in algebra and higher mathematics.

Finance: In finance, exponents are used to calculate the growth of investments and the performance of investment portfolios. Annualized growth rates are commonly expressed through exponents.

You are correct; in finance, exponents are essential for calculating and understanding the growth of investments and the performance of investment portfolios.

Investment Growth: Exponents are fundamental in the formula for compound interest. When you invest money in a compound interest account, your earnings or account balance grow exponentially over time. The general formula for calculating compound interest is $A = P(1 + r/n)^{(nt)}$, where "A" is the final balance, "P" is the principal (initial amount), "r" is the annual interest rate, "n" is the number of times interest is compounded per year, and "t" is the time in years. Here, the exponents (n and nt) express exponential growth.

Portfolio Performance: Investors often use annualized rates of return or compound growth rates to assess the performance of their investment portfolios over time. These rates are expressed in terms of exponents and help determine how much an investment has grown or declined in value over a given period.

Present and Future Value: Exponents are also applied in calculating net present value (NPV), which is a metric used to evaluate the viability of investment projects. NPV involves discounting future cash flows using discount rates, which are often expressed in exponential form.

Growth Rates and Returns: In finance, growth rates, such as sales growth rate or revenue growth rate, are expressed in terms of exponents to indicate the exponential increase of a financial metric over time. This is important for assessing a company's financial health and growth potential.

Understanding Returns: Exponents help investors and financial analysts understand and compare the performance of different investments or assets over time. The use of exponential rates allows capturing the cumulative effect of gains or losses.

In summary, exponents are a fundamental tool in finance for calculating and expressing the growth, return, and present value of investments and projects. These exponential applications are crucial for making informed decisions in the world of finance and investments.

Technology: Digital electronic circuits rely on operations with binary exponents. In emerging technologies, such as quantum computing, operations on qubits also involve exponential calculations.

Exponents have important applications in the field of technology, especially in areas such as digital electronics and emerging technologies like quantum computing. Here are more details on how exponents are applied in these contexts:

Digital Electronic Circuits: Digital electronic systems use binary representations, where data is stored and processed in the form of bits (0 and 1). Arithmetic operations, such as multiplication and exponentiation, are fundamental in the operations of an arithmetic logic unit (ALU) of a processor. Operations with binary exponents are essential for calculating powers and executing algorithms that involve exponential operations.

Quantum Computing: In quantum computing, calculations are performed using qubits instead of classical bits. Qubits have the property of being in multiple states at once, allowing for exponential calculations to be performed much more efficiently than in classical computers. Some quantum algorithms, such as Shor's algorithm, stand out for their ability to factor large numbers, a task that would be extremely slow on a conventional computer.

Quantum Cryptography: Quantum cryptography is a field that relies on the principles of quantum computing to develop highly robust security systems.

Quantum cryptography algorithms use exponential calculations and quantum properties to ensure security in communications and transactions.

Signal Processing: In applications of signal processing, such as data compression and image and audio processing, exponents are used in fast Fourier transform (FFT) algorithms and discrete cosine transform (DCT) algorithms. These algorithms are fundamental for digital signal processing and are applied across a wide range of technological applications, from multimedia file compression to data communication.

Machine Learning and Neural Networks: In the field of machine learning, deep learning algorithms often involve exponential calculations. These models are essential for tasks such as pattern recognition, natural language processing, and computer vision.

Exponents are essential for a variety of technological applications, from classic digital electronics to emerging technologies like quantum computing. These exponential applications allow for complex calculations and problem-solving in technology efficiently and effectively. They are a fundamental mathematical tool for modeling and understanding a wide range of phenomena in natural sciences, social sciences, and technological applications. They allow for the description of growth, rates of change, probabilities, system dynamics, and much more, making them a fundamental concept in problem-solving across numerous fields.

23.Square Roots: Finding the number that, when multiplied by itself, gives another number

The square root is a mathematical operation that involves finding a number which, when multiplied by itself, produces a given number. This number is called the "square root" and is represented by the symbol "√".

Some important characteristics of square roots include:

Notation: The square root of a number "x" is denoted as √x. For example, the square root of 9 is written as √9 and equals 3, since 3 × 3 = 9. The notation for the square root of a number "x" uses the symbol "√" followed by the number. In your example, the square root of 9 is written as "√9" and equals 3, since 3 multiplied by itself (3 × 3) equals 9. This illustrates that the square root of a number is the value that, when squared, yields the original number. It is the inverse operation of squaring a number.

This notation is widely used in mathematics and science to indicate the square root operation and to express the relationship between a number and its square root. For example:

- √16 = 4, because 4 × 4 = 16.
- √25 = 5, because 5 × 5 = 25.

The square root notation is essential for understanding and working with this concept in various mathematical and scientific applications.

Positive and Negative Square Roots: In most contexts, when the "square root" of a number is mentioned, it refers to the positive square root. However, each number actually has two square roots: one positive and one negative. For example, the square root of 9 is both 3 and -3, since both 3 × 3 and -3 × -3 equal 9.

The term "square root" generally refers to the positive root in most contexts. However, as mentioned, each number has two square roots: one positive and one negative.

For example, in the case of the square root of 9, you have both the positive square root, which is 3 (because 3 × 3 = 9), and the negative square root, which is -3 (because -3 × -3 = 9). Both square roots are mathematically valid and satisfy the property that, when multiplied by themselves, they produce the original number.

In certain contexts—especially in mathematics and equations—it is important to consider both the positive and negative square roots, since both may be valid solutions. For example, when solving a quadratic equation, it is common to find two solutions, one positive and one negative, corresponding to the two square roots of the discriminant.

It is important to keep this in mind when working with square roots in various mathematical and scientific applications.

Negative Square Root Notation: To specifically represent the negative square root of a number, a minus sign is placed in front of the square root symbol, as in $-\sqrt{x}$. This clearly indicates that you are considering the negative square root, which is the negative value that, when multiplied by itself, produces the number "x".

In mathematics, this notation is important when working with equations or contexts that require both the positive and negative square roots, as previously mentioned.

Real and Complex Numbers: While the square roots of positive numbers are real numbers, negative numbers do not have real square roots, because no real number squared gives a negative result. Instead, the square roots of negative numbers are complex numbers.

In mathematics, square roots of positive numbers are real numbers, meaning you can find a real number that, when squared, gives the positive number. However, negative numbers do not have real square roots, because there is no real number that, when squared, gives a negative number. Instead, the square roots of negative numbers are complex numbers.

Complex numbers are an extension of real numbers that include imaginary components. The square root of a negative number is represented as a complex number and usually expressed with a real part and an imaginary part. For example, the square root of -9 is 3i, where "i" represents the imaginary unit. This means that 3i is the square root of -9, since $(3i)^2 = -9$.

Complex numbers play an important role in mathematics and have various applications, including in physics and engineering.

Property of the Square Root: The square root of a number "x" is a number "y" such that $y^2 = x$. Therefore, to find the square root of a number, you are looking for the number that, when squared, produces the original value.

For example, to find the square root of 9, you are looking for a number "y" such that $y^2 = 9$. In this case, y equals 3, since $3^2 = 9$, which satisfies the square root property.

This property is essential in mathematics and is used in numerous contexts, such as in geometry to find the length of a side of a square given its area, or in physics to determine the velocity of an object based on its kinetic energy. The square root is also the inverse operation of squaring, which makes it useful for undoing squared values.

Perfect Square Roots: Some numbers have exact square roots that are integers. For example, the square root of 4 is 2, the square root of 25 is 5, and the square root of 100 is 10. These numbers are considered "perfect square roots."

Correct—numbers that have exact square roots which are integers are called perfect square roots. This means that when you calculate the square root of numbers like 4, 25, or 100, you get an integer as a result.

For example:

- The square root of 4 is 2, because 2 × 2 = 4.
- The square root of 25 is 5, because 5 × 5 = 25.
- The square root of 100 is 10, because 10 × 10 = 100.

These perfect square roots are useful in mathematics and in various applications because they simplify calculations and allow for easier representation of numbers. However, most numbers are not perfect square roots, and in those cases, the square root is represented as a decimal or fractional number.

Perfect square roots are helpful because they provide whole-number values that simplify calculations and make representation more manageable. For example, if you need to calculate the length of a side of a square with an area of 25 square units, you know that the square root of 25 is 5, so the side length is 5 units.

However, in most cases, numbers are not perfect squares, meaning their square roots are decimal or fractional numbers. In such situations, numerical approximations are used—such as rounding to a specific decimal —or the square root is represented as a fraction. This is common in math, physics, engineering, and other disciplines that require precise calculations.

Calculators and Software: In practice, square roots are commonly calculated using calculators, math software, or built-in math functions in spreadsheets and programming languages.

Calculators and mathematical software have greatly simplified the calculation of square roots, especially for numbers with many decimal places or complex fractions. Most scientific calculators and mathematical software applications allow for easy calculation of square roots, and some even offer the ability to work with complex numbers when the square root involves a negative number. This facilitates a wide range of mathematical and scientific computations.

Applications: Square roots are a fundamental part of mathematics and are applied in various areas—from geometry to physics and engineering—to calculate lengths, areas, and volumes. They are also useful in fields such as algebra and calculus to solve equations and mathematical problems.

Square roots are an essential concept in mathematics and science due to their versatility and applicability across a wide range of fields. In physics, they are used to describe quantities such as velocity and acceleration, as well as to calculate the lengths of sides in geometric shapes. In engineering, they are fundamental to the design and analysis of structures, electrical systems, and mechanical systems. In statistics, they are used to calculate standard deviations and to understand variability in data.

Their usefulness extends to economics, biology, chemistry, and many other disciplines, making them a valuable tool for problem-solving and understanding real-world phenomena.

24.Probability: The chance that something will happen, usually expressed as a fraction or percentage

Probability refers to the measure of the likelihood that a particular event or outcome will occur. It is commonly expressed as a fraction or a percentage, where 0 represents the impossibility of the event occurring, and 1 (or 100% in percentage terms) represents certainty that it will happen. The calculation of probability involves considering the number of favorable outcomes (cases where the event occurs) in relation to the total number of possible outcomes.

For example, if you roll a fair six-sided die, the probability of getting a "6" is 1/6 or approximately 16.67% (1/6 * 100%).

Probability theory is an important branch of mathematics that is used in a wide range of applications, from gambling to decision-making in science, business, and many other areas. It helps to quantify uncertainty and make decisions based on an understanding of the possibilities and risks involved in an event or situation.

Probability refers to the likelihood of an event occurring and is typically expressed as a fraction or a percentage. It is a measure that varies between 0 (indicating that the event is impossible) and 1 (indicating that the event is certain), or in percentage terms, between 0% and 100%. Probability is a fundamental part of probability theory and is used in a wide variety of fields, such as statistics, mathematics, science, business, and decision-making, to assess and manage uncertainty.

Probability is a fundamental tool in probability theory and plays a crucial role in evaluating and managing uncertainty across a wide range of contexts and disciplines. Some examples of areas where probability is significantly applied include:

Statistics: Probability is used to analyze data and make inferences about populations from samples. In statistics, probability plays a fundamental role in analyzing data and making inferences about populations based on samples.

Probability distributions: These are used to describe how the possible values of a random variable are distributed. Examples include the normal distribution, the binomial distribution, and the Poisson distribution.

Central limit theorem: This theorem states that, under certain conditions, the distribution of the mean of a random sample from a population approaches a normal distribution, regardless of the shape of the original distribution.

Statistical inference: Probability is used in statistical inference to make claims about a population based on data from a sample. This includes estimating population parameters and conducting hypothesis tests.

Confidence intervals: Confidence intervals are based on probability and are used to estimate a plausible range of values for a population parameter.

Hypothesis testing: Hypothesis tests are based on probability and are used to determine whether there is sufficient evidence in the data to reject or fail to reject a null hypothesis.

Random sampling: Probability is used to ensure that sample selection is unbiased and representative of the population, which is essential for making valid inferences.

In summary, probability is an essential component of statistics and is employed to understand and quantify uncertainty in data and to make informed decisions about populations from limited samples.

Gambling: In casinos and betting games, probability is used to calculate the odds of winning or losing in various games. Probability plays a crucial role in gambling, such as in casinos and other betting games. In these environments, probability is used to calculate the odds of winning or losing in various games, helping players make informed decisions and establishments manage risk.

Roulette: In roulette, probability distributions are used to calculate the odds of the ball landing on a particular number or color. Players can use these probabilities to make decisions about their bets.

Blackjack: In blackjack, probability is employed to determine when it is more likely that a high or low card will be dealt, which can influence players' decisions about whether to hit or stand.

Poker: In poker, probability is used to calculate the odds of getting certain hands (such as a straight or flush) based on the dealt cards. Players can use this information to make strategic decisions during the game.

Slot machines: Slot machines are designed with algorithms that use probabilities to determine when winning combinations will occur and how much will be paid out to players.

Sports betting: In sports betting, probability is used to calculate the odds that a team will win a game, influencing the odds offered to bettors.

Understanding probability in gambling is essential for both players and operators, as it allows for the assessment of risk and advantage in a particular game. Players can make more informed decisions about how and when to bet, and casinos can set profit margins and manage their operations more effectively.

Natural sciences: In physics, chemistry, and biology, probability is employed to describe random phenomena and model complex systems. Probability is significantly used in the natural sciences, including physics, chemistry, and biology, to describe random phenomena and model complex systems.

Physics: In quantum physics, probability is essential for describing the behavior of subatomic particles, such as electrons and photons. Quantum mechanics uses wave functions that represent the probability of finding a particle in a specific location. Additionally, thermodynamics relies on probability to describe the statistical behavior of particles in macroscopic systems, such as gases.

Chemistry: In chemistry, probability is used to model the distribution of electrons in atoms and to predict the likelihood of certain chemical reactions occurring. Chemical kinetics also relies on probabilistic models to describe how the concentrations of reactants and products change over time.

Biology: Biology employs probability in a wide range of applications. For example, in genetics, it is used to predict the probability of inheriting certain genetic traits. In ecology, probabilistic models are used to understand population dynamics and interactions within ecosystems. In epidemiology, probability is used to model the spread of diseases.

Probability is also utilized in many other branches of the natural sciences to address the inherent uncertainty of complex systems and phenomena. It helps scientists quantify randomness and make informed decisions about how to model and understand the natural world.

Finance: Probability is used in risk management, investments, and the evaluation of financial assets. In finance, probability plays a fundamental role in risk management, investments, and the evaluation of financial assets.

Risk assessment: Probability is used to quantify and assess risk in financial decisions. This includes calculating the likelihood of adverse events occurring, such as losses in investments or defaults on debts, which is essential for financial risk management.

Value at Risk (VaR) models: In finance, VaR is a measure that uses probability to estimate how much an investment portfolio could lose in a specific period with a given level of confidence. Investors and asset managers use VaR to assess risk exposure.

Investments: Investors use probability to make investment decisions. They assess the odds of gains or losses in financial assets, such as stocks, bonds, funds, and other investment instruments, before making buy or sell decisions.

Option and derivative evaluation: Probability is used to assess the value and risk associated with options and other financial derivatives. Models like the Black-Scholes model employ probability to calculate option prices and exercise probabilities.

Financial planning and insurance: In personal and business financial planning, probability models are used to estimate future financial positions and to determine the amount of insurance needed to cover financial risks, such as health, property, and life.

Portfolio analysis: Fund managers and investors use probability techniques to build and evaluate diversified investment portfolios and to estimate the performance and risk of those portfolios.

Probability is a critical tool in the world of finance for assessing risks, making informed investment decisions, and managing financial assets efficiently. It helps financial professionals understand and quantify uncertainty in markets and make data-driven decisions based on probabilistic analysis.

Decision-making: In business and management, probability is used to make informed decisions by considering different scenarios and their probabilities. In the realm of decision-making in business and management, probability plays a crucial role by helping leaders and managers make informed decisions when considering different scenarios and the probabilities associated with those scenarios. Some ways in which probability is applied in decision-making in business include:

Risk assessment: Probability is used to assess the risks associated with various business decisions, such as investments, product launches, geographical expansion, etc. This enables companies to identify and mitigate potential threats.

Strategic planning: In strategic planning, organizations can use scenario analysis involving different possible situations. Probability is used to assign probabilities to these scenarios and evaluate their impact on strategic goals and objectives.

Supply chain management: Companies use probability to forecast product demand, evaluate delivery times, and manage inventory efficiently.

Project evaluation: Probability is employed in project evaluation, calculating net present value (NPV) and other financial indicators that depend on future cash flows and their associated probabilities.

Marketing decisions: Marketing strategies often involve evaluating the probabilities of success in advertising campaigns, product launches, and market segmentation strategies.

Human resources: Probability is used in human resources management to predict the likelihood of candidates' success in selection processes and to calculate the risk associated with hiring, promotion, and employee retention decisions.

The application of probability in business decision-making helps reduce uncertainty by providing a quantitative basis for evaluating options and considering different possible outcomes. This allows organizations to make more informed and strategic decisions, optimizing resources and minimizing risks.

Engineering: In engineering, probability is applied to assess the reliability of systems and design robust solutions. In the field of engineering, probability is applied in various ways, especially to assess the reliability of systems and design robust solutions.

System reliability: Probability is used to assess the reliability of systems and components. Engineers can calculate the likelihood of failure of different parts of a system and use this information to design more resilient systems and implement preventive maintenance strategies.

Risk analysis: In engineering projects, risk analyses are conducted to assess the probabilities and consequences of undesirable events, such as structural failures or accidents. This is crucial in the construction, civil engineering, and aerospace industries, among others.

Experimental design: In engineering, experiments are often used to test and validate designs. Probability is used to design experiments that are statistically sound, allowing for meaningful results with a minimal number of tests.

Quality control: Probability is fundamental in quality control, where it is used to assess the likelihood that a product meets certain specifications. Control charts and control limits are based on probabilistic techniques.

Software reliability: In software engineering, probability is used to assess the reliability and security of systems and applications. This is essential to ensure that software functions correctly and is resilient to failures.

Traffic and transportation engineering: Probability is employed to model traffic flow, evaluate road capacities and transportation systems, and predict congestion and travel times.

The application of probability in engineering contributes to data-driven decision-making, improves system reliability, and reduces risks. It helps engineers design more effective and safe solutions and optimize the performance and lifespan of products and systems.

Social sciences: In sociology and political science, probability is used to conduct surveys and public opinion studies. In the social sciences, including areas such as sociology and political science, probability is used to carry out surveys and public opinion studies.

Opinion surveys: Probability is used to design surveys that are representative of the population. This involves randomly selecting individuals or groups to form samples that are statistically significant and accurately reflect the opinions and attitudes of the general population.

Probabilistic sampling: In conducting surveys, probabilistic sampling is used to ensure that each member of the population has a known chance of being selected for the sample. This helps avoid biases and ensures the validity of the results.

Statistical models: Probability is used in constructing statistical models to analyze survey data and other studies. This includes estimating confidence intervals for results, identifying relationships between variables, and evaluating statistical significance.

Election predictions: In political science, probability is applied to predict election outcomes and electoral surveys. Probabilistic models may take into account survey data, historical trends, and other factors to forecast the outcome of an election.

Social research: Probability is also applied in broader social science research to assess the likelihood of certain social events or behaviors occurring and to analyze uncertainty related to social theories and concepts.

The use of probability in sociology and political science is essential to ensure that the data collected is representative and that the conclusions drawn from studies are reliable and valid. This allows researchers and surveyors to better understand public opinion, social behaviors, and political processes, and to make decisions based on more solid data.

Probability is an essential tool that helps us understand and quantify uncertainty across a wide range of fields, enabling more informed decision-making and more accurate analysis in situations where the outcome is uncertain or random.

25.Statistics: The collection and analysis of data

Statistics focuses on the collection, analysis, and interpretation of data. It is a fundamental discipline in many areas of science, research, and decision-making, playing a crucial role in understanding patterns, trends, and relationships in data. Here is a breakdown of the main activities related to statistics:

Data Collection: The process of statistics begins with the collection of data. This involves gathering relevant information from diverse sources, which may include surveys, experiments, observations, historical records, or other data sources.

Selection of Data Sources: To conduct statistical research, it is important to determine the appropriate data sources. These sources can vary depending on the context and purpose of the research. They may include surveys, experiments, observations, historical records, existing databases, sensors, questionnaires, interviews, among others.

Design of Data Collection: Once the data source is selected, the data collection process must be designed. This involves determining how the data will be obtained, what variables will be measured, and how they will be recorded. The design of data collection should be as rigorous and representative as possible to ensure the validity of the results.

Data Collection: This stage involves executing the data collection plan. Researchers or surveyors may conduct surveys, carry out experiments, collect data from existing records, observe events, or engage in other activities according to the previously established design.

Data Recording and Storage: The collected data is recorded and stored in an organized and secure manner. This may involve entering data into databases, spreadsheets, or data management systems.

Data Validation: A validation process is carried out to ensure that the collected data is accurate and reliable. This may include reviewing data for errors, verifying the consistency of records, and correcting any identified issues.

Data collection is a critical phase in the statistical process, as the quality and integrity of the data are fundamental to the accuracy and validity of subsequent analyses. Any errors at this stage can have a significant impact on the final results, making careful planning and execution necessary.

Data Organization: Once the data is collected, it is organized and recorded in a structured manner. This may involve creating tables, graphs, databases, or other representations to facilitate subsequent analysis. Data organization is a fundamental step following data collection in the statistical process.

Data Structuring: The collected data often comes from various sources and may be in different formats. At this stage, it is organized coherently and structured to be more manageable and analyzable. This may involve converting data into a standard format.

Creation of Databases: In many cases, data is stored in databases, which allow for efficient access, management, and analysis of the data. Databases can be powerful tools for organizing large datasets.

Data Tabulation: Tables are created to represent the data, especially when dealing with categorical data or data with multiple variables. Tables can help visualize information clearly and structured.

Graphs and Visualizations: In addition to tables, graphs and visualizations are used to represent data in a more visual and comprehensible manner. This may include bar graphs, pie charts, scatter plots, and other types of graphs.

Data Sorting: In some cases, data is sorted according to certain variables, which can be useful for identifying patterns and trends more easily.

Creation of Derived Variables: Sometimes, new variables are generated from the original data, which can be useful for conducting specific analyses. These derived variables may include rates, proportions, averages, among others.

Data organization is essential for facilitating the analysis and interpretation of data. A structured and orderly presentation of data allows researchers and analysts to better understand the information, identify patterns, perform statistical calculations, and make data-driven decisions.

Data Summary: Statistics involves summarizing data concisely using measures of central tendency (such as mean, median, and mode) and measures of dispersion (such as standard deviation and interquartile range) to describe key characteristics of the data. Data summarization is an essential part of statistical analysis used to describe and understand datasets.

Measures of Central Tendency:

Mean: The arithmetic mean is the average of the values in a dataset. It is calculated by summing all values and dividing by the number of observations.

Median: The median is the value that lies at the center of an ordered dataset. It divides the data into two equal halves.

Mode: The mode is the value that appears most frequently in a dataset. There may be one mode (unimodal data) or multiple modes (bimodal or multimodal data).

Measures of Dispersion:

Standard Deviation: The standard deviation measures the dispersion of data relative to the mean. It indicates how much the data tends to vary from the average value.

Interquartile Range: The interquartile range is the difference between the third quartile (Q3) and the first quartile (Q1). It helps assess variability in the central portion of the data, excluding extreme values.

Percentiles: Percentiles divide a dataset into one hundred equal parts. The 50th percentile is the median, the 25th percentile is the first quartile, and the 75th percentile is the third quartile. Percentiles are useful for understanding data distribution and identifying outliers.

Histograms: Histograms are graphs that represent data distribution in intervals. They help visualize the shape of the distribution and the concentration of data in different ranges.

Box Plots: Box plots are graphs that show data distribution by representing quartiles, outliers, and the median. They are useful for comparing distributions and detecting extreme values.

Data summarization is essential for understanding the structure and key characteristics of a dataset. These measures and techniques provide a concise description that helps analysts and decision-makers identify patterns, trends, and outliers in the data.

Data Analysis: At this stage, statistical techniques are applied to examine the data in greater detail. This may include hypothesis testing, regression analysis, analysis of variance, among other methods, depending on the research objectives. Data analysis is a critical phase in the statistical process where various statistical techniques are applied to gain a deeper and more meaningful understanding of the collected data.

Hypothesis Testing: Hypothesis tests are used to evaluate claims or assumptions about the data. For example, tests can be performed to determine whether there is a significant difference between two groups or to verify if an observed relationship is statistically significant.

Regression Analysis: Regression analysis is used to examine the relationship between a dependent variable and one or more independent variables. It can be simple (one independent variable) or multiple (multiple independent variables) and allows predicting values of the dependent variable based on the independent variables.

Analysis of Variance (ANOVA): ANOVA is used to compare the means of three or more groups to determine if there are significant differences among them. It can help identify which group or groups are different from the others.

Time Series Analysis: Time series analysis is applied when data is collected over time. Statistical techniques are used to identify patterns, trends, and seasonality in time series data.

Principal Component Analysis: This technique is used to reduce the dimensionality of data by identifying the most important variables in a multivariate dataset. It helps simplify the interpretation of complex data.

Cluster Analysis: Cluster analysis is used to group similar observations into sets or clusters. It is useful in data segmentation and identifying clustering patterns.

Correlation Analysis: Correlation analysis is used to measure the relationship between two variables. The correlation coefficient indicates the strength and direction of the relationship between the variables.

Survival Analysis: In certain studies, such as medical research, survival analysis is used to assess the likelihood of an event occurring within a specific time period.

Data analysis is essential for extracting meaningful insights from the collected data and addressing specific research questions. Appropriate statistical techniques are chosen based on research objectives and the nature of the data. Moreover, statistical analysis aids in making data-driven decisions and generating conclusions supported by statistical evidence.

Data Interpretation: The results of the analysis are interpreted to draw meaningful conclusions. The aim is to understand the relationships, patterns, or differences that may arise from the data and determine their relevance to the issue at hand.

Data interpretation is the final phase of the statistical process, where meaningful conclusions are drawn from the results of data analysis. Here are the key aspects of data interpretation:

Understanding the Results: Before making interpretations, it is crucial to thoroughly understand the results of the analysis. This involves reviewing statistical outputs, graphs, and summary measures to identify patterns, trends, and relationships in the data.

Relation to Research Objectives: Data interpretation should align with the research objectives. Conclusions that are relevant to answering research questions or addressing the issue at hand need to be identified.

Identifying Relationships and Patterns: The aim is to identify significant relationships between variables and patterns in the data. This may include cause-and-effect relationships, correlations, trends over time, or differences between groups.

Evaluating Statistical Significance: It is important to determine whether the observed relationships and differences are statistically significant. This is done through hypothesis testing or other appropriate statistical techniques.

Presentation of Findings: The interpreted results are presented clearly and effectively through reports, presentations, or data visualizations. Communicating findings is essential for others to understand and utilize the information.

Generating Conclusions and Recommendations: Data interpretation should lead to the generation of evidence-based conclusions. These conclusions can support recommendations for decision-making or specific actions.

Considering Uncertainty: Interpretation should take into account the inherent uncertainty in the data and statistical analysis. It is important to recognize the limitations of the results and communicate the reliability of the conclusions.

Practical Application: Conclusions derived from data interpretation should translate into practical actions or decisions. This may involve implementing changes, formulating policies, conducting further research, or taking corrective actions.

Data interpretation serves as the bridge between analysis results and informed decision-making. It is a critical stage where valuable insights are extracted from the data and their relevance is determined for solving problems, advancing research, or improving processes.

Presentation of Results: Results are communicated through reports, graphs, tables, and presentations. Effective presentation of findings is essential for stakeholders to make informed decisions.

The presentation of results is a crucial part of the statistical process, as findings must be communicated effectively for interested parties to understand them and make informed decisions.

Choice of Formats: Results can be presented in various formats, including written reports, graphs, tables, oral presentations, online formats, interactive visualizations, among others. The choice of format depends on the target audience and communication objectives.

Clarity and Conciseness: The presentation of results should be clear and concise. Avoid unnecessary technical jargon and use language that is understandable to the audience. Complex statistical concepts should be explained simply.

Effective Visualizations: Graphs and tables play a fundamental role in presenting results. They should be designed effectively, using appropriate colors, labels, and titles to highlight key information.

Data Stories: Instead of merely presenting numbers or isolated results, it is helpful to tell a "data story" that explains the context, findings, and implications. This helps make sense of the results.

Summary of Conclusions: It is important to summarize the main conclusions and findings at the beginning of the presentation so that the audience has a quick understanding of the most relevant points.

Evidence and Support: Ensure that the presented results are supported by solid evidence, including specific data and relevant statistical analyses.

Target Audience: Consider who the recipients of the presentation will be and tailor the content and tone according to their needs and level of understanding.

Interaction: If possible, allow interaction with the results. This may include exploring interactive visualizations or the opportunity to ask questions in real-time during an oral presentation.

Visual Resources: Utilize visual resources, such as graphs, images, and concrete examples, to help illustrate and support key points.

Ethics and Privacy: Ensure compliance with ethical standards and privacy when presenting data, especially if it involves personal or sensitive information.

Effective presentation of results is essential for decision-makers, colleagues, and other stakeholders to understand findings and take data-driven actions. A good presentation not only informs but also persuades and motivates action.

Decision Making: Statistical results are used to support decision-making across a wide variety of fields, from business and government to scientific research and planning.

Decision-making supported by statistical results is a fundamental process in a wide range of fields and is used to address problems, improve processes, and guide actions.

Business and Companies:

Strategic Management: Companies use statistical analysis to assess market trends, competition, and economic factors, allowing them to make strategic decisions, such as expansion, product diversification, or entry into new markets.

Resource Management: The management of human, financial, and operational resources benefits from statistical analysis for making decisions about hiring, resource allocation, and budgeting.

Process Optimization: Companies use statistics to identify inefficiencies in production and operational processes and take measures to optimize them.

Government and Politics:

Policy Formulation: Governments use statistical data to understand the needs of the population and design effective public policies in areas such as health, education, security, and the environment.

Urban Planning: City and regional planning is based on statistical data to make decisions about development, transportation, and land use.

Elections and Voting: Survey results and statistical analysis influence political decisions and campaign strategies.

Scientific Research:

Hypothesis Validation: Statistical results are used to validate or refute scientific hypotheses. Scientists make decisions about accepting or rejecting their hypotheses based on statistical evidence.

Experimental Design: Planning and conducting scientific experiments involve making decisions about which variables to measure, sample sizes, and statistical methods.

Health and Medicine:

Diagnosis and Treatment: Doctors and healthcare professionals use statistical data to make decisions about diagnosis and treatment, as well as to evaluate the effectiveness of therapies and treatments.

Clinical Research: Clinical trials and medical research studies rely on statistical analysis to determine the safety and efficacy of new drugs and treatments.

Education:

Performance Evaluation: Schools and universities use statistical data to evaluate the performance of students and teachers, influencing decisions about academic programs and resources.

Educational Planning: Educational policies are based on statistical data to make decisions about resource allocation, curriculum planning, and evaluation of educational systems.

Environment and Sustainability:

Natural Resource Management: Statistical data is used to make decisions about managing natural resources, such as water, energy, and forests, to ensure long-term sustainability.

Finance:

Investments: Investors use statistical analysis to make decisions about asset allocation and investment selection.

Risk Management: Companies and financial institutions use statistical risk analysis to make decisions about insurance, loans, and risk coverage.

Statistical results provide objective and quantitative evidence that supports decision-making in all these fields and many others. They enable decision-makers to consider multiple scenarios, assess risks and opportunities, and select the most informed and beneficial option.

Statistics is a powerful tool for understanding the world and answering questions from data. It helps identify trends, assess the significance of results, estimate populations from samples, and support evidence-based decision-making. It is an essential discipline in the information age.

26.Angles: Measures of the opening between two lines at a common point

Angles are measures of opening between two lines that intersect at a common point, known as the vertex of the angle. Angles are fundamental in geometry and are used to describe spatial relationships and directions. Here are some key concepts related to angles:

Vertex: The common point where the two lines forming an angle meet is called the vertex of the angle.

Sides: The two lines extending from the vertex are the sides of the angle. Each side extends in a different direction from the vertex.

Amplitude: The amplitude of an angle is the measure of its opening, which is generally expressed in degrees (°), minutes ('), and seconds ("). A complete angle is equivalent to 360 degrees.

Classification of Angles:

Acute Angle: An angle with an amplitude of less than 90 degrees.

Right Angle: An angle with an amplitude of exactly 90 degrees.

Obtuse Angle: An angle with an amplitude greater than 90 degrees but less than 180 degrees.

Straight Angle: An angle with an amplitude of 180 degrees, which is a completely flat angle.

Complete Angle: An angle with an amplitude of 360 degrees, which is equivalent to a full circle.

Complementary Angles: Two angles are called complementary if the sum of their amplitudes equals 90 degrees. For example, a 30-degree angle and a 60-degree angle are complementary. Complementary angles are two angles whose amplitudes sum to 90 degrees. This property is fundamental in geometry and in solving problems related to angles. In other words, if you have two complementary angles, one will be acute (with an amplitude of less than 90 degrees) and the other will be obtuse (with an amplitude greater than 90 degrees).

For example, if you have a 30-degree angle and a 60-degree angle and you add them together, you get: 30 degrees + 60 degrees = 90 degrees This demonstrates that they are complementary angles, as their sum equals 90 degrees. This property is used in many geometric and mathematical problems to determine unknown angles when the relationship of complementarity is known.

Supplementary Angles: Two angles are called supplementary if the sum of their amplitudes equals 180 degrees. For example, a 120-degree angle and a 60-degree angle are supplementary. Supplementary angles are two angles whose amplitudes sum to 180 degrees. This property is another fundamental concept in geometry and is opposite to the idea of complementary angles. In other words, if you have two supplementary angles, one will be acute (with an amplitude of less than 90 degrees) and the other will be obtuse (with an amplitude greater than 90 degrees).

For example, if you have a 120-degree angle and a 60-degree angle and you add them together, you get: 120 degrees + 60 degrees = 180 degrees This demonstrates that they are supplementary angles, as their sum equals 180 degrees. Similar to complementary angles, this property is used in solving geometric and mathematical problems to determine unknown angles when the relationship of supplementarity is known.

Vertically Opposite Angles: When two lines cross, the angles opposite each other are equal. This means that if two angles have the same vertex and share a side, their amplitudes are equal. Vertically opposite angles are an important concept in geometry. When two lines intersect (forming an intersection), four angles are created at the intersection point. The angles opposite each other at the intersection and sharing the same vertex are called vertically opposite angles. The key property of vertically opposite angles is that they have the same amplitude. This property is known as the "Vertically Opposite Angles Theorem" and is expressed as follows: "Vertically opposite angles are equal." In mathematical terms, if α and β are two vertically opposite angles, then: $\alpha = \beta$ This property is useful in solving geometric problems and in proving theorems in geometry. It also applies in situations where it is necessary to determine the measure of an unknown angle by leveraging the equality of vertically opposite angles with known angles.

Adjacent Angles: Adjacent angles are those that share a side and a vertex. The sum of their amplitudes is equal to the amplitude of the angle formed by their other common side. Adjacent angles are two angles that share a side and a vertex. The key property of adjacent angles is that the sum of their amplitudes is equal to the amplitude of the angle formed by their other common side. This property is commonly used in geometry and in solving problems related to angles.

In mathematical terms, if α and β are two adjacent angles, and they share the same vertex and the same side, then: $\alpha + \beta = \gamma$ Where γ is the angle formed by the common side of α and β. This property is useful for calculating the amplitude of unknown angles when the relationship of adjacency between known angles is understood. It is also used in the proof of theorems and in solving geometric problems where multiple angles are involved in a geometric figure.

Angles are fundamental concepts in mathematics and science and are used in various fields, from geometry and trigonometry to physics and engineering, to describe spatial relationships and solve geometric problems.

Geometry: In geometry, the properties and relationships of angles are studied, including concepts such as vertically opposite angles, adjacent angles, and complementary angles. Geometry also uses angles to analyze shapes, calculate areas and perimeters, and solve problems related to location and position. Angles are a fundamental component in geometry and play a key role in describing geometric figures, as well as analyzing their properties and relationships. Here are some more detailed applications of angles in geometry:

Angles in Polygons: In polygon geometry, the concepts of angles are used to describe the measures of the internal and external angles of regular and irregular polygons. This is essential for calculating areas and perimeters of polygons.

Triangle Interior Angles Theorem: One of the most important theorems in geometry states that the sum of the internal angles of a triangle is always equal to 180 degrees. This theorem is based on the property of the internal angles of a triangle.

Congruence of Triangles: In triangle congruence, angles play a central role. Two triangles are considered congruent if their corresponding sides and angles are equal.

Pythagorean Theorem: In the Pythagorean theorem, which is fundamental in geometry, right angles (of 90 degrees) in right triangles are worked with. This theorem relates the lengths of the sides of a right triangle.

Angles in the Cartesian Plane: In the Cartesian plane, angles are used to describe the orientation of lines and segments, which is important in analytical geometry and in the graphical representation of functions.

Angles in Geometric Transformations: Angles are critical in the study of geometric transformations, such as rotations, reflections, and translations.

Determining how angles change during these transformations is essential for understanding their effects on geometric figures.

Angles are a fundamental component of geometry used to describe and analyze geometric figures, calculate measures, prove theorems, and deeply understand geometric properties. Their study is essential in mathematics and science and has practical applications across various fields, from architecture to engineering and cartography.

Trigonometry: Trigonometry focuses on the relationships between angles and sides of triangles. Angles play a central role in trigonometric functions, such as sine, cosine, and tangent, which are used to solve problems related to angle measures and distances. Trigonometry is a branch of mathematics that focuses on the study of the relationships between angles and sides of triangles, and angles are fundamental in trigonometric functions.

Triangles and Angles: Triangles are geometric figures that involve angles and sides. In trigonometry, angles in triangles play a central role, especially in right triangles, where one of the angles is a right angle of 90 degrees.

Basic Trigonometric Functions: The three basic trigonometric functions are sine (sin), cosine (cos), and tangent (tan). These functions are defined in terms of the sides and angles of a triangle and are used to relate measures of angles and distances in trigonometric problems.

Solving Triangles: Trigonometry is used to solve triangles, meaning you can find measures of unknown angles and sides if you know certain relationships between them. This is useful in a variety of applications, such as navigation, physics, and engineering.

Inverse Trigonometric Functions: In addition to the basic trigonometric functions, trigonometry includes inverse trigonometric functions, such as arcsine (asin), arccosine (acos), and arctangent (atan). These functions are used to find angles from trigonometric relationships.

Trigonometric Identities: Trigonometric identities are equations that relate trigonometric functions to each other. These identities are useful for simplifying trigonometric expressions and solving trigonometric equations.

Angle Representation: In trigonometry, angles are typically measured in radians, an angular measure that is fundamental in more advanced calculations. A radian is the measure of the central angle of a circle that subtends an arc equal to its radius.

In summary, angles play an essential role in trigonometry, which is a valuable mathematical tool for solving problems related to measures of angles, distances, and relationships in triangles and circles. Trigonometry has applications in fields ranging from navigation to theoretical physics and is used to model and solve a wide range of phenomena in science and engineering.

Physics: In physics, angles are essential for describing movements, directions, and spatial relationships. The concepts of angles are applied in fields such as kinematics, mechanics, and optics to analyze trajectories, forces, and the reflection and refraction of light. In physics, angles are fundamental to describing and analyzing a variety of phenomena and concepts related to motion, direction, and spatial relationships.

Kinematics: In kinematics, which is the branch of physics that deals with the motion of objects, angles are used to describe the trajectories and velocities of moving objects. Angular velocity and acceleration are measured in radians per second, and angles are essential for describing the direction of motion.

Mechanics: In mechanics, which studies the behavior of objects in response to forces, angles are used to describe the orientation of forces and moments. The concepts of angles are fundamental for understanding how forces influence the rotation and equilibrium of objects.

Optics: In optics, which deals with the study of light and its behavior, angles are crucial for describing the reflection and refraction of light on surfaces and through various media. The angles of incidence and refraction are fundamental for understanding how light behaves when passing from one medium to another.

Thermodynamics: In thermodynamics, which focuses on the study of energy and heat, angles may be related to the orientation of thermal systems and devices like engines. The concepts of angles are also used in the context of rotating thermodynamic systems.

Electromagnetism: In electromagnetism, which studies the interactions between electric charges and magnetic fields, angles are used to describe the direction and orientation of electric and magnetic fields in space.

Quantum Mechanics: In quantum mechanics, which is the fundamental theory that describes the behavior of subatomic particles, the concepts of angles are fundamental for describing angular momentum and properties of particles like electrons and atoms.

In all these areas of physics, angles play an essential role by providing a way to describe and measure spatial and directional properties, allowing for a deeper understanding of physical phenomena and the resolution of problems related to motion, forces, and the interaction of light and matter.

Engineering: In engineering, angle measurements are used to design and analyze structures, mechanical systems, and electrical circuits. Precision in angle measurement is critical to ensure that components and systems function correctly. In engineering, precision in measurement and the use of angle measurements are critical for the effective design, analysis, and functioning of a wide range of components and systems. Here are some key applications of angle measurements in engineering:

Structural Design: In civil and structural engineering, angle measurements are used to design and analyze the integrity of structures such as bridges, buildings, dams, and towers. Accurate determination of angles is fundamental to ensuring the stability and safety of these structures.

Mechanical and Machine Engineering: Measurement and control of angles are essential in mechanics for designing mechanical components such as gears, cams, shafts, and motion transmission systems. Mechanical systems require precise alignment of angles for efficient operation.

Electrical and Electronic Engineering: In electrical and electronic engineering, angle measurements are important for designing circuits and devices. In the manufacturing of integrated circuits and electronic components, precision in angle alignment is critical for the functioning of devices like microchips and sensors.

Aerospace Engineering: Aerospace engineering uses angle measurements in the design and control of spacecraft, aircraft, and navigation systems. Precision in angle measurement is vital for flight control and spatial navigation.

Robotics: In robotics, accurate measurement of angles is fundamental for controlling robots and robotic arms. Angle sensors and motion control systems are used to ensure that robots move and operate accurately.

Manufacturing Engineering: In manufacturing, precise angle measurements are required for the alignment and assembly of mechanical, electronic, and structural components. Precision in angle alignment ensures that manufactured products meet specifications.

Naval Engineering: In the construction of ships and vessels, angle measurement is fundamental to ensure stability and efficiency in navigation.

The precise measurement of angles and their application in engineering is essential to ensure the quality, safety, and performance of products and systems across a variety of fields. Engineers rely on reliable and accurate angle measurements to successfully carry out design and construction projects.

Surveying and Cartography: In surveying and cartography, angles are used to measure and represent the location of points on Earth. This is essential for creating maps and planning construction and land development projects. In surveying and cartography, angles play a critical role in the precise measurement and representation of the location of points on Earth.

Topographic Surveys: Surveyors use instruments such as theodolites and total stations to measure horizontal and vertical angles between points on the ground. These angles are used to determine coordinates and elevations of points, which is essential for creating detailed topographic maps.

Azimuth Angles: In surveying, azimuth refers to the horizontal angle between a reference direction (usually north) and a line of observation. Azimuth angles are used to determine the direction of lines and features on the ground.

Polar Coordinates: In some surveying systems, polar coordinates are used that include distances and angles to describe the location of points in relation to a reference point or baseline.

Digital Cartography: In the modern era, Geographic Information Systems (GIS) and digital cartography use data on angles and distances to create accurate maps and navigation systems. GPS (Global Positioning System) relies on angle measurements to determine the exact location of a receiver on Earth.

Map Drawing: Cartographers use angles to plot lines of latitude and longitude on maps. They also use angles to represent geographical features and to create cartographic projections that allow the curved surface of the Earth to be represented on a plane.

Construction Project Planning: Civil engineers and architects use angle measurements in surveying to plan construction projects. This includes determining the location and orientation of buildings, roads, and other infrastructures.

Geodetic Mapping: Geodesy is the science that deals with measuring and modeling the shape of the Earth. Geodesists use precise angle measurements to determine the shape and size of the Earth, as well as to perform high-precision geodetic measurements.

Surveying and cartography are disciplines that heavily rely on angular measurements and geographic coordinates to represent and analyze the Earth's surface. Precision in angle measurement is essential to ensure that maps and geospatial data are reliable and useful in a variety of applications, from construction project planning to navigation and land management.

Astronomy: Astronomy uses angles to measure the position and motion of celestial bodies in the sky. Angles are used to determine the location of stars, planets, comets, and other astronomical objects. In astronomy, angles play a crucial role in measuring and describing the position and motion of celestial bodies in the sky.

Celestial Coordinates: Celestial coordinates, which include right ascension and declination, are used to specify the position of celestial objects in the sky. These coordinates are based on angle measurements and provide a standardized way to locate stars, planets, and other objects on the celestial sphere.

Determination of Local Time: Sky angles are used to determine local time based on the apparent position of the Sun or stars. This is essential for astronomical navigation and clock synchronization.

Observations and Telescopes: Astronomers use telescopes equipped with altazimuth or equatorial mounts to track and aim at celestial objects accurately. These telescopes move at angles to follow the rotation of the Earth and to point at objects in the sky.

Distance Measurement: Angles are used in combination with stellar parallax and other methods to measure distances to stars and other astronomical objects. This is fundamental for determining the scale of distances in the universe.

Study of Planetary Movements: Observing the angles and positions of planets over time has been essential for understanding their orbits and movements in the solar system.

Space Navigation: Space missions, such as probes and space telescopes, use precise angular measurements to accurately point and observe specific objects and regions in space.

Astrophotography: Astrophotography involves capturing images of celestial objects using cameras and telescopes. Angles are essential for pointing and framing images.

Astronomy is a science that relies on precise angular measurements to observe, document, and understand the universe. Precision in angle measurement is fundamental for tracking celestial objects and conducting advanced astronomical research, such as determining the position and motion of stars, planets, galaxies, and other distant objects in space.

Navigation: Navigation, whether on land or at sea, involves measuring and calculating angles to determine location and direction. Angles are essential in the compass, GPS, and other navigation instruments. Navigation, both on land and at sea, heavily depends on measuring and calculating angles to determine location and direction. Angles play a fundamental role in orientation and positioning in navigation.

Compass: A compass is a navigation instrument that uses a magnetic pointer to indicate the north direction. Navigators use the compass to measure angles relative to magnetic north and to determine the direction they are moving.

GPS (Global Positioning System): GPS relies on a network of satellites that transmit signals to receivers on Earth. GPS receivers calculate their exact location by measuring the arrival angles of signals from multiple satellites. These angles are used to determine the geographic coordinates (latitude and longitude) of a receiver.

Sextant: The sextant is an instrument used in maritime navigation to measure the height of celestial bodies, such as the Sun and stars, above the horizon. These angular measurements are used to calculate the latitude and longitude of a vessel at sea.

Nautical Charts: Nautical charts are detailed maps designed for maritime navigation. These charts often include course lines, which are lines that show the direction of a reference point in relation to the vessel at a certain angle.

Dead Reckoning: Dead reckoning is a navigation method that involves tracking the direction and speed of a vessel over time using navigation logs. This includes measuring angles to calculate course and drift.

Piloting: In coastal navigation, navigators may use visual references, such as lighthouses and buoys, to take bearings and reference angles to avoid obstacles and stay on a safe course.

Land Navigation: In land navigation, such as hiking orientation or route planning in aerial navigation, angles are used to determine direction and location.

Measuring and calculating angles are essential in navigation for determining direction, location, and accurately tracking routes. Navigation has greatly benefited from technological advances such as GPS, but angular concepts remain fundamental to understanding and using modern navigation tools.

The concepts of angles are fundamental in a wide range of disciplines and are applied to describe spatial relationships, measure directions, and solve geometric problems. Their understanding and use are essential in mathematics and science, as well as in practical applications in everyday life and various professional fields.

27.Coordinates: Points on a plane represented by ordered pairs (x, y)

Coordinates are used to represent points on a plane and are generally expressed as ordered pairs (x, y). Coordinates are fundamental in mathematics, physics, geometry, and many other disciplines to describe and locate points in a two-dimensional space.

Geometry: In geometry, coordinates are used to describe the location of points, plot lines and segments, calculate areas and perimeters of figures, and solve problems related to geometric shapes on a Cartesian plane.

Representation of Points: Cartesian coordinates are used to represent the location of points on the plane. A point is represented as an ordered pair (x, y), where "x" is the coordinate on the horizontal axis (abscissa) and "y" is the coordinate on the vertical axis (ordinate).

Trajectories and Segments: Coordinates are used to plot straight lines and line segments between two points on the Cartesian plane. This is fundamental for describing the relationships of distance and direction between points.

Calculating Distances: Coordinates are used to calculate the distance between two points on the plane, applying the Pythagorean theorem or other distance formulas, depending on the situation.

Areas of Figures: Coordinates are used to calculate the areas of geometric figures, such as triangles, quadrilaterals, and irregular polygons. This is achieved by dividing the figure into segments and calculating the areas of individual regions.

Perimeters of Figures: Coordinates are used to calculate the perimeters of figures by summing the lengths of the sides. This is important in the design and construction of structures and objects.

Geometric Transformations: Coordinates are also used in geometric transformations, such as translations, rotations, symmetries, and reflections. These transformations change the coordinates of the points, which is fundamental in geometry.

Equations of Lines and Curves: Equations in the Cartesian plane are used to represent straight lines, curves, and circles. These equations allow for the description and analysis of the geometry of these shapes.

Solving Geometric Problems: Coordinates are used to solve a wide variety of geometric problems, from determining intersections and points of tangency to calculating shaded areas in complex figures.

Cartesian coordinates are an essential tool in geometry for describing and analyzing geometric figures, calculating distances and areas, and solving problems related to shapes and geometric properties in a two-dimensional plane. Additionally, coordinates are used in analytic geometry to carry out more advanced investigations.

Physics: In physics, coordinates are used to describe the position and movement of objects in a two-dimensional space. This is fundamental in kinematics and the study of particle mechanics.

Kinematics: Kinematics focuses on the study of the motion of objects without considering the causes of motion. Coordinates are used to describe the position of an object as a function of time. Two-dimensional coordinates (such as (x, y)) are used to represent the trajectory of an object on the plane.

Classical Mechanics: Classical mechanics, which includes Newtonian mechanics, relies on the use of coordinates to describe the motion of particles and systems of particles. Motion equations, such as Newton's equations and parabolic motion equations, are expressed in terms of spatial coordinates.

Study of Trajectories: Coordinates are used to study the trajectories of moving objects. This is essential for understanding how objects move in space and how they interact with each other.

Particle Physics: In particle physics, coordinates are used to describe the position and direction of subatomic particles in particle accelerators and high-energy experiments.

Analysis of Projectile Motion: Coordinates are used to analyze the motion of projectiles, such as the launching of objects in a gravitational field. This is essential in applications such as ballistics and astronautics.

Dynamics of Rotating Bodies: In the study of rotating objects, such as wheels and gears, angular coordinates (angles) are used to describe the position and angular velocity of these objects.

Waves and Optics: In the study of waves and optical phenomena, coordinates are used to describe the propagation of waves and the interaction of light with objects and surfaces.

Space Navigation: In space and aerospace navigation, coordinates are used to describe the position and movement of spacecraft and satellites in space.

Celestial Mechanics: Celestial mechanics deals with the study of the motion of celestial bodies in space. Coordinates are essential to describe the orbits of planets, stars, and other celestial objects.

Coordinates play a fundamental role in physics by allowing for the description and analysis of the motion and position of objects in two-dimensional space. These coordinates are essential for formulating motion equations, solving physics problems, and understanding a wide range of physical phenomena in the real world.

Coordinate Systems: Different coordinate systems, such as Cartesian coordinates, polar coordinates, and spherical coordinates, are used to describe points in space based on their own characteristics and angular relationships.

Coordinate systems are fundamental tools in mathematics and science for describing the position and spatial relationships between points in three-dimensional space. Each coordinate system has its own characteristics and is used based on the simplicity and applicability of the situation.

Cartesian Coordinates (Rectangular): This is the most common coordinate system and is based on perpendicular axes. In a three-dimensional Cartesian coordinate system, three orthogonal axes are used: the x-axis, the y-axis, and the z-axis. The coordinates of a point are represented as a triplet (x, y, z), where "x" is the coordinate on the x-axis, "y" is the coordinate on the y-axis, and "z" is the coordinate on the z-axis. This system is widely used in Euclidean geometry and in various applications in physics and mathematics.

Polar Coordinates: In polar coordinates, the position of a point is described in terms of its distance from an origin (r) and the angle (θ) it makes with respect to a reference axis. This system is often used in situations where distance and angle are more relevant than Cartesian coordinates. It is common in applications such as the representation of circular and conic movements.

Spherical Coordinates: Spherical coordinates are used to describe the position of a point in terms of its radial distance (r), its polar angle (θ), and its azimuthal angle (φ). This system is commonly used to describe locations in three-dimensional space and is particularly useful in applications such as celestial mechanics and particle physics, where spherical angles simplify the description of the direction of objects in space.

Each of these coordinate systems has its own advantages and is used in specific situations depending on what is most convenient for representing the objects or phenomena under study. The choice of the appropriate coordinate system depends on the nature of the problem and the simplicity it allows in describing spatial relationships between points.

Geography: In geography, geographic coordinates (latitude and longitude) are used to represent the location of places on Earth. This is essential for navigation and cartography.

In geography, geographic coordinates are fundamental for accurately representing the location of places on Earth. Geographic coordinates, which include latitude and longitude, allow geographers and cartographers to describe and locate any point on the Earth's surface.

Latitude: Latitude refers to the imaginary lines running east to west around the Earth. It is measured in degrees, minutes, and seconds north (north latitude) or south (south latitude) of the equator, which is the latitude line at 0°. Latitude varies from 0° at the equator to 90° at the North and South Poles.

Longitude: Longitude refers to the imaginary lines running north to south around the Earth. It is measured in degrees, minutes, and seconds east (east longitude) or west (west longitude) of the Greenwich Meridian, which is the longitude line at 0°. Longitude varies from -180° to +180°.

Geographic coordinates are expressed in degrees, minutes, and seconds, allowing for an accurate representation of the location of any point on Earth. For example, New York City is located at approximately 40° 42' north latitude and 74° 0' west longitude.

The applications of geographic coordinates in geography include:

Navigation: Geographic coordinates are fundamental for land and maritime navigation. Navigation devices use latitude and longitude to determine the exact location of a place and plot routes.

Cartography: Maps and topographic charts use geographic coordinates to accurately represent the location of geographic features, cities, roads, and more.

Geographic Information Systems (GIS): GIS are powerful tools that use geographic coordinates to manage and analyze geospatial data, which is essential in urban planning, resource management, and many other applications.

Climatic and Geological Studies: The precise location of weather stations, observatories, and geological events is described using geographic coordinates.

Natural Resource Localization: Geographic coordinates are used to locate and manage natural resources such as mines, oil wells, and conservation areas.

Earth and Environmental Studies: Geographers use geographic coordinates to conduct research on climate change, biodiversity, and environmental impacts.

Tourism and Travel: Geographic coordinates are useful for travel planning, identifying destinations, and navigating unfamiliar environments.

Geographic coordinates are an essential tool in geography and play a critical role in representing and locating places on Earth, facilitating navigation, cartography, and a wide range of geospatial studies.

Navigation: In navigation, coordinates are used to determine the position of a vessel, aircraft, or vehicle on a map or navigation system.

In navigation, coordinates are used to pinpoint the exact position of a vessel, an aircraft, or any other vehicle on a map or navigation system. Precise location determination is essential for safe and effective navigation.

Maritime Navigation: Mariners use geographic coordinates, such as latitude and longitude, to determine the position of a ship on the ocean. Positioning systems like GPS provide precise coordinates that allow navigators to know exactly where their vessel is at sea.

Aerial Navigation: In aviation, geographic coordinates are essential for plotting flight paths, identifying landmarks, and maintaining precise control over the position of an aircraft in the sky. Air navigation systems, such as GPS and inertial navigation systems, provide precise coordinates for aerial navigation.

Land Navigation: In land navigation, whether in terrestrial vehicles like cars and trains or outdoor activities like hiking, coordinates are used to determine location and plan routes. GPS devices and mapping applications on smartphones are common tools for land navigation.

Navigation Instruments: In all forms of navigation, navigation instruments such as compasses, sextants, GPS, and navigation charts are used to calculate and record the coordinates of their position.

Route Planning: Route planning involves the use of coordinates to design and plot optimal routes, avoiding obstacles and hazards in navigation. This is essential in travel planning and in navigating vessels and aircraft.

Emergency Navigation: In emergency situations, such as sea rescues or emergency landings, coordinates are vital for rescue services to accurately locate people in need and provide assistance.

Coordinates play an essential role in navigation, whether at sea, in the air, or on land. They provide navigators with the ability to know their exact position and plot safe and efficient routes, which is critical for safety and effectiveness in navigation in all its forms.

Technical Drawing and Engineering: In technical drawing and engineering, coordinates are used to represent and specify the location of objects and components in plans and designs.

In technical drawing and engineering, the use of coordinates is fundamental to represent and specify the precise location of objects, components, and features in plans and designs. This is essential for effective communication in the creation, manufacturing, and construction of products, buildings, and systems.

Coordinate Systems: Technical drawings and plans are based on coordinate systems that establish a common reference for all elements of the design. Cartesian coordinate systems, such as the XYZ coordinate system, are common in three-dimensional technical drawing.

Location of Points and Lines: Coordinates are used to specify the location of points of interest on a plan, as well as the position of lines and segments that define shapes and structures.

Dimensioning: Dimensions are expressed in terms of coordinates to indicate the size and position of objects and components in a drawing. This allows manufacturers to accurately understand the required measurements.

Reference to Origin Coordinates: Plans often include an origin point with known coordinates from which all other locations and dimensions in the drawing are measured.

Tolerance Specifications: Coordinates are also used in tolerance specifications, where allowable variations in dimensions and position of elements are indicated.

Computer-Aided Design (CAD): CAD programs use coordinate systems and drawing tools that allow designers and engineers to create and modify designs accurately and efficiently.

Georeferencing: In applications such as Geographic Information System (GIS) design and urban planning, geographic coordinates are used to locate features on the Earth's surface.

Three-Dimensional Modeling: In 3D modeling, three-dimensional coordinate systems are used to represent objects in three-dimensional space. This is relevant in fields such as 3D printing and product engineering.

Representation of Structures: Coordinates are used to represent the location and orientation of components in structures, such as buildings and machines.

Technical Documentation: Technical drawings and plans are a form of technical documentation that uses coordinates to provide clear and detailed information about the construction, assembly, or operation of a product or system.

The use of coordinates in technical drawing and engineering allows for precise and universal communication in the representation of objects and systems. This is fundamental to ensuring that designs can be manufactured or constructed accurately and meet the required specifications.

Mathematics: Coordinates are used in a variety of mathematical fields, such as linear algebra, calculus, and analytical geometry, to solve equations and perform calculations.

Coordinates are a fundamental tool in mathematics and are used in various mathematical fields to perform calculations, solve equations, and study geometric relationships and properties.

Linear Algebra: In linear algebra, coordinates are used to represent vectors and matrices. Vectors can be described in terms of coordinates in a vector space. Operations with vectors, such as addition and scalar multiplication, are performed using coordinates. Matrices are also represented by coordinates and are used to solve systems of linear equations.

Analytical Geometry: Analytical geometry combines geometry with algebraic techniques and uses coordinates to study geometric properties and spatial relationships. Cartesian coordinates are common in analytical geometry, where points in a plane are represented as ordered pairs (x, y). This allows the representation and study of lines, curves, and geometric figures through equations and calculations.

Calculus: In calculus, coordinates are used to describe and analyze mathematical functions. Functions can be represented by equations in terms of coordinates, such as $f(x) = y$, where "x" and "y" are the coordinates. Calculus allows the calculation of derivatives and integrals of functions using techniques based on coordinates.

Differential Geometry: Differential geometry is a mathematical field that focuses on the study of curves and surfaces in three-dimensional spaces. Coordinates are used to describe and analyze these structures, and partial derivatives and other mathematical concepts are applied in this context.

Multilinear Algebra: In multilinear algebra, coordinates are used to work with vector spaces of more than three dimensions. This field relates to the study of tensors and differential forms, and coordinates are fundamental for expressing and manipulating these mathematical entities.

Projective Geometry: Projective geometry is a field that focuses on the study of properties that remain invariant under projective transformations. In this context, homogeneous coordinates are used to describe points and lines, allowing for an elegant and uniform representation of projective geometry.

Fractal Geometry: In fractal geometry, coordinates are used to define and study fractal sets, which are highly irregular and self-similar geometric structures. Coordinates are essential for constructing and analyzing fractals.

Topology: In topology, coordinates are used to describe and analyze topological spaces and apply topological concepts such as continuity and convergence.

Coordinates are a versatile and powerful tool in mathematics, used across a wide range of mathematical fields to represent objects, functions, and structures, facilitating the resolution of equations, the study of geometric properties, and the analysis of mathematical relationships in various dimensions and contexts.

Computer Programming: In programming, coordinates are used to define the position of graphical elements on a computer screen. This is important in the development of games, graphical applications, and user interfaces.

In computer programming, especially in software development involving graphical elements, coordinates are essential to define the position and arrangement of objects on a computer screen. This is fundamental in creating games, graphical applications, and user interfaces.

2D Graphics: In game development, drawing applications, and other programs that work in two dimensions, coordinates are used to specify the location of graphical elements, such as characters, objects, buttons, and text. Cartesian coordinates (x, y) are commonly used to represent position in the 2D plane.

3D Graphics: In three-dimensional environments, such as video games and 3D modeling applications, three-dimensional coordinate systems (x, y, z) are used to define the position and orientation of objects in 3D space. 3D coordinates allow for the creation of virtual worlds and the simulation of realistic three-dimensional environments.

User Interfaces (UI): In user interface design, coordinates are used to locate interface elements, such as buttons, windows, drop-down menus, and dialog boxes. This ensures that elements are displayed correctly on the screen and are interactive.

Mouse and Touch Events: In user interface programming, coordinates are used to detect and respond to user input events, such as mouse clicks or touches on a touchscreen. The coordinates of the mouse pointer or touch are used to determine the location of the event.

Animation: Coordinates are essential in the animation of graphical elements. By modifying coordinates over time, animations of objects on the screen can be achieved, such as smooth movements, transitions, and visual effects.

Collisions: In games and simulations, coordinates are used to detect collisions between objects. By comparing the coordinates of different objects, it can be determined if they have collided and take appropriate action.

Screen and World Coordinates: Coordinate systems for screens and the world are often used. Screen coordinates are used to draw objects on the screen, while world coordinates represent the actual position of objects in the context of the program.

2D/3D Graphics Development: In game and graphics development environments, libraries and frameworks are provided that simplify coordinate handling, object representation, and graphical manipulation.

Coordinates play a crucial role in computer programming when it comes to creating graphical elements, user interfaces, and games. Proper manipulation of coordinates allows developers to control the position and interaction of objects on the screen, which is essential for creating visual applications and interactive experiences.

Whether in the form of ordered pairs (x, y) in a two-dimensional plane or in more complex coordinate systems in three-dimensional spaces, coordinates are a fundamental tool for representation, analysis, and description of positions and spatial relationships across a wide variety of disciplines and applications.

28.Patterns: Logical sequences of numbers or figures

Patterns refer to logical sequences of numbers, figures, objects, or events that follow a predictable rule or structure. These patterns can be visual, numerical, or conceptual and can be found in a variety of contexts. Patterns are fundamental in mathematics, science, design, and many other disciplines.

Numerical Patterns: In mathematics, numerical patterns are sequences of numbers that follow a specific rule. For example, the sequence 2, 4, 6, 8, 10 follows an arithmetic pattern in which each number is 2 units greater than the previous one. Prime numbers are another example of a numerical pattern, as they follow a specific rule in terms of divisibility. Numerical patterns are sequences of numbers that follow specific and predictable rules. These patterns can take various forms and be of different types, and they are fundamental in mathematics for understanding and predicting numerical sequences.

Arithmetic Pattern: In an arithmetic pattern, each number in the sequence is obtained by adding (or subtracting) a fixed amount called the "difference" to the previous number. For example, the sequence 3, 6, 9, 12 follows an arithmetic pattern with a difference of 3.

Geometric Pattern: In a geometric pattern, each number in the sequence is obtained by multiplying (or dividing) the previous number by a fixed amount called the "ratio." For example, the sequence 2, 4, 8, 16 follows a geometric pattern with a ratio of 2.

Perfect Squares Pattern: Perfect squares, such as 1, 4, 9, 16, 25, follow a specific pattern. Each number is the square of the corresponding natural number (1^2, 2^2, 3^2, 4^2, 5^2).

Even/Odd Numbers Pattern: Sequences of even numbers (2, 4, 6, 8, ...) and odd numbers (1, 3, 5, 7, ...) are common numerical patterns. Even numbers are multiples of 2, while odd numbers are not divisible by 2.

Fibonacci Series: The Fibonacci series is a numerical pattern in which each number is the sum of the two preceding numbers in the sequence: 0, 1, 1, 2, 3, 5, 8, 13, 21, 34, ...

Triangular Numbers: Triangular numbers follow a pattern in which each number is the sum of consecutive natural numbers. For example, 1, 3, 6, 10 are triangular numbers.

Powers of 2 Sequence: The powers of 2 sequence follows a numerical pattern in which each number is 2 raised to an increasing power: 1, 2, 4, 8, 16, 32, ...

Prime Numbers: Prime numbers are a special type of numerical pattern in which each number is divisible only by 1 and itself. Examples of prime numbers include 2, 3, 5, 7, 11, 13, 17, 19, 23, ...

Recognizing numerical patterns is an important skill in mathematics and can help predict future values, find solutions to mathematical problems, and understand relationships between numbers. These patterns are fundamental in many areas of mathematics and have applications in science, engineering, and computer science, among other fields.

Geometric Patterns: In geometry and design, geometric patterns are sequences of figures or shapes that follow a predictable structure. For example, a geometric pattern could consist of a series of concentric circles of different sizes or a sequence of equilateral triangles interspersed with squares. Geometric patterns are sequences of figures or shapes that follow a predictable structure and are used in geometry, design, and various creative applications to create interesting and visually appealing compositions. These patterns can be simple or complex and are based on the repetition or variation of geometric elements.

Geometric Mosaics: Geometric mosaics are patterns created by repeating geometric shapes, such as squares, triangles, or hexagons, in such a way that they fill a surface uniformly and form a harmonious design.

Fractals: Fractals are geometric patterns that repeat at different scales. A famous example is the Mandelbrot set, which is generated through iterations of simple mathematical equations and creates highly detailed fractal patterns.

Tile Patterns: In architecture and interior design, geometric tile patterns are used to create designs on floors, walls, and ceilings. These patterns can range from simple designs of squares and rectangles to more intricate designs with complex geometric shapes.

Textile Design: Geometric patterns are used in textiles, such as fabrics and rugs, to create attractive designs. Patterns may include repetitions of circles, diamonds, lines, and other geometric elements.

Wallpaper Design: In interior design, geometric patterns are used in wallpaper to add visual interest to walls. These patterns can range from simple stripes and checks to more intricate and abstract designs.

Logo Design: Logos and brands often incorporate geometric patterns to create a distinctive visual identity. Patterns can be an important part of a brand's aesthetics and recognition.

Abstract Art: Artists use geometric patterns in abstract art to create unique visual compositions. These patterns can range from simple repetitive shapes to more complex and abstract designs.

Game Design: In video game and board game design, geometric patterns are used to create maps, boards, and visual elements that are both attractive and functional.

Geometric patterns are an effective way to create visually appealing and coherent designs. They can convey a sense of order, symmetry, and balance, or they can be used to create more dynamic and abstract designs. Geometry and geometric patterns are an integral part of creativity in various artistic and design fields.

Sequence Patterns: Sequence patterns are found in many fields, including music, where a sequence of notes may follow a predictable melodic or rhythmic pattern. In computer science, scripts follow logical execution patterns that determine the flow of a program.

Sequence patterns are logical sequences of events, elements, or actions that follow a predictable structure and are found in diverse fields, including music, computer science, and others. These patterns are essential for music composition, computer programming, and in many other contexts.

Music: In music, sequence patterns are fundamental for composing melodic and rhythmic pieces. Melodic patterns may include sequences of notes or chords that repeat coherently throughout a song. Rhythmic patterns determine the duration and rhythm of notes and are used to create the rhythmic structure of a musical composition.

Computer Science and Programming: In computer science, sequence patterns are crucial for programming the logic of software. Scripts and programs follow logical execution patterns that include the sequence of instructions, decision-making (through conditional control structures), and repetition (through loops). Programming uses patterns to define the structure and behavior of applications and systems.

Genetic Sequences: In biology, genetic sequences follow specific patterns. For example, DNA contains sequences of nucleotides that determine genetic information and the structure of genes. Identifying patterns in genetic sequences is crucial for research in genomics and molecular biology.

Signal Processing: In signal processing, such as in telecommunications, audio and video signals follow sequence patterns. Detecting patterns in signals is important for the transmission and reception of information.

Linguistics and Natural Language: In linguistics and natural language processing, sequence patterns are used to analyze and understand human language. Linguistic patterns may include sequences of words or grammar that follow predictable rules.

Process Control: In automation and industrial process control, sequence patterns are used to define and monitor the operations of machinery and systems. Sequence patterns are essential for ensuring efficiency and safety in industrial production.

Economics and Finance: In economics and finance, sequence patterns can be identified in market movements, economic trends, and financial data. These patterns help in making investment decisions and economic predictions.

Navigation and Routes: In navigation systems, such as GPS, sequence patterns are used to determine routes and directions. Sequence patterns of streets, roads, or geographic coordinates are fundamental for accurate navigation.

In summary, sequence patterns are found in a wide variety of fields and disciplines, and they are essential for the organization, prediction, and control of events and processes. Identifying and understanding these patterns is fundamental for advancement in many areas of study and application.

Natural Patterns: Patterns are found in nature, such as the growth sequence of leaves on a plant or the arrangement of scales on a reptile's skin. These natural patterns often follow mathematical and geometric rules.

Natural patterns are sequences and structures found in nature that follow mathematical and geometric rules. These patterns can be observed in a wide variety of phenomena and organisms in nature. Often, these natural patterns have been an area of interest for scientists, mathematicians, and artists due to their beauty and complexity.

Fibonacci in Nature: The Fibonacci sequence (1, 1, 2, 3, 5, 8, 13, 21, 34, ...) is found in many aspects of nature. For example, spirals in snail shells and the arrangement of leaves on a plant often follow patterns based on the Fibonacci sequence.

Fractals in Nature: Fractals, geometric patterns that repeat at different scales, can be found in nature. For example, the branching patterns of trees and shrubs often follow a fractal structure.

Animal Skin: The arrangement of scales on the skin of reptiles, such as snakes, follows specific geometric patterns. Additionally, patterns on the skin of animals, such as leopards or zebras, have been studied for their camouflage capabilities.

Patterns in Crystals: Crystals have highly ordered atomic structures that follow mathematical and geometric patterns. The way atoms group in crystals can result in patterns of fractals or specific symmetries.

Patterns in Water Currents: Water currents, such as rivers and streams, often follow meandering and bifurcation patterns that adhere to geometric and dynamic rules.

Beach Formations: Patterns created by ocean waves on the sand of a beach often follow specific geometric patterns, such as concentric waves.

Symmetry in Organisms: Symmetry is found in many organisms, from bilateral symmetry in animals such as fish and humans to radial symmetry in organisms like starfish.

Patterns in Plant Growth: Plant growth follows specific patterns, such as the spiral arrangement of leaves that conforms to the Fibonacci sequence or the formation of fractals in roots and branches.

Patterns in the Sky: Natural patterns are also found in the sky, such as the arrangement of stars in constellations and the formation of galaxies that follow laws of physics and mathematics.

These examples illustrate how mathematical and geometric patterns manifest in nature in surprising ways. The study of these patterns is not only scientifically fascinating but has also inspired artists and designers throughout history.

Color and Design Patterns: In graphic design and fashion, color and shape patterns are used to create visually appealing designs. Patterns can be symmetrical, asymmetrical, repetitive, or random, depending on the designer's intent.

Color and design patterns are fundamental elements in the world of graphic design, fashion, and other creative disciplines. These patterns are created through the repetition or variation of colors, shapes, and textures with the aim of achieving attractive and coherent visual designs.

Graphic Design: In graphic design, color and design patterns are used to create backgrounds, illustrations, logos, and other visual elements in projects such as websites, prints, advertising, and promotional materials. Patterns can be symmetrical, asymmetrical, or repetitive, depending on the style and intent of the design.

Fashion: In fashion, design patterns are applied to fabrics and garments to create unique and attractive clothing designs. These patterns can include geometric, floral, abstract designs, and more. Fashion designers consider the scale, repetition, and harmony of color patterns to achieve a desired look.

Interior Design: In interior design, patterns are used in wallpaper, fabrics, rugs, and other decorative elements to add texture and style to a space. Patterns can influence the atmosphere of a room and reflect the personality of the owner.

Product Design: Design patterns are applied to products, from dishware and textiles to electronics and packaging. Patterns can be an important part of a product's visual identity and aesthetic appeal.

Art and Illustration: Artists and illustrators use color and design patterns in their works to create striking visual compositions. These patterns can be part of an artist's distinctive style.

Stationery Design: Patterns are used in creating stationery, greeting cards, wrapping paper, and other related products. Patterns can be seasonal, thematic, or abstract.

Website and Application Design: In web and application design, color and design patterns are applied to elements such as buttons, backgrounds, and page layouts. Patterns help maintain visual consistency and guide user interaction.

Textile and Print Design: In the textile industry, weaving and printing patterns are created and applied to fabrics for garment making, curtains, and bedding. Patterns can vary in scale and style, from stripes and checks to more elaborate designs.

The choice of color and design patterns is a creative and strategic decision that can influence how a design or product is perceived. Designers carefully consider how patterns interact with other visual elements, such as typography and images, to achieve an attractive and effective visual outcome.

Behavior Patterns: In psychology and sociology, human behavior patterns can be identified in social or individual situations. These patterns are sometimes used to predict future behaviors or understand social dynamics.

Behavior patterns are systematic observations of how people behave in social or individual situations over time. These patterns are of interest in fields such as psychology, sociology, and other disciplines related to social sciences. Identifying and understanding human behavior patterns is fundamental for predicting future behaviors, making informed decisions, and understanding social dynamics. Here are examples of how behavior patterns are applied in various areas:

Psychology: In psychology, behavior patterns are studied to understand how individuals respond to stimuli, situations, and events. Behavior patterns are investigated in areas such as child development, clinical psychology, cognitive psychology, and social psychology. These patterns can help identify psychological disorders, assess the effects of therapeutic interventions, and understand decision-making.

Sociology: Sociology focuses on the study of society and human behavior in broader social contexts. Sociologists investigate behavior patterns in social groups, communities, and cultures to understand issues such as social stratification, group dynamics, social movements, and cultural changes.

Behavioral Economics: In behavioral economics, patterns of economic decision-making are analyzed. Behaviors related to spending, saving, investing, and consumer choice are studied to understand why individuals make certain financial decisions.

Marketing and Advertising: In marketing and advertising, consumer behavior patterns are used to develop effective strategies. Analyzing online consumer behavior data, such as click tracking and purchases, is used to tailor advertising campaigns and marketing strategies.

Education: In the field of education, student behavior patterns are observed to assess academic performance and design more effective teaching methods. Data on student behavior, such as attendance rates and exam performance, can be used to identify areas for improvement.

Security and Public Policy: Behavior patterns are also important in public policy formulation and security. They can be used to analyze crime trends, drug consumption patterns, traffic behaviors, among others, to make informed decisions and improve public safety.

Technology and Data Science: In the digital age, large amounts of data about online human behavior are collected. Data scientists and analysts use this data to identify behavior patterns, such as search preferences, social media interactions, and online purchasing patterns.

Human Resources: In human resources management, employee behavior patterns are observed to assess performance, job satisfaction, and team dynamics. These patterns can guide decision-making related to hiring, training, and professional development.

Human behavior patterns are fundamental in a variety of fields, and their analysis and understanding significantly impact decision-making, policy formulation, and improvements in quality of life in society. Research and analysis of these patterns are essential for advancing the understanding of the human mind and social dynamics.

Patterns in Data: In data analysis, patterns are sought in data sets to identify trends, relationships, or anomalies. This is fundamental in data science and data-driven decision-making.

Analyzing patterns in data is an essential component of data science and data-driven decision-making. It involves identifying, interpreting, and understanding patterns, trends, relationships, and anomalies in data sets. This practice is fundamental in various areas, from business analysis to scientific research.

Business Analysis: Companies use pattern analysis in data to understand customer behavior, identify market trends, optimize internal processes, and make strategic decisions. This may include sales analysis, customer segmentation, fraud detection, and demand forecasting.

Data Science: In data science, professionals explore data to uncover patterns that may serve as the basis for predictive models. This includes techniques such as machine learning and data mining to identify relationships and patterns in large data sets.

Medicine: In medical research, patient data is analyzed to identify patterns that can help prevent, diagnose, and treat diseases. This includes analyzing clinical data, medical imaging, and genetic data.

Finance: In the financial sector, patterns in data are used for fraud detection, credit risk analysis, market movement prediction, and investment portfolio management.

Meteorology: Climate scientists analyze meteorological data to identify patterns that help predict weather and extreme weather events, such as hurricanes and tornadoes.

Scientific Research: In various scientific disciplines, from physics to biology, experimental data is analyzed to identify patterns that support scientific theories and allow for new research.

Education: In the field of education, student performance data is analyzed to identify patterns that indicate areas for improvement in teaching and learning.

Government: Government agencies use pattern analysis in data to understand demographic trends, make public policy decisions, and manage resources efficiently.

Security: In security, video, audio, and network data are analyzed to identify patterns that may indicate threats or anomalous behaviors.

Digital Marketing: In digital marketing, user online behavior data, such as clicks and purchases, is analyzed to personalize campaigns and improve user experience.

Analyzing patterns in data involves using statistical, mathematical, and computational tools and techniques to reveal valuable insights. It helps organizations and professionals make more informed decisions, anticipate future events, and identify areas for focus and improvement. Moreover, it is an essential part of data science and data-driven decision-making in the digital age.

Patterns are an important way to simplify and understand information across a variety of fields. They allow for the identification of rules, the prediction of outcomes, and the creation of visually appealing designs. The ability to recognize and work with patterns is a valuable skill in mathematics, science, design, and many other disciplines.

29.Logic: Reasoning based on rules and relationships to reach conclusions

Logic is a fundamental discipline that focuses on reasoning based on rules and relationships with the purpose of reaching valid conclusions. Essentially, logic is concerned with coherence and validity in the thought process. Here are some key characteristics of logic:

Rules and Principles: Logic is based on a set of rules, principles, and structures that guide reasoning. These rules are essential for determining whether an argument is valid or not. The rules and principles are fundamental in the study of logic. They establish the guidelines for valid and coherent reasoning.

Principle of Identity: This principle states that anything is identical to itself. In symbolic terms, A is equal to A. Therefore, if something is true, it is true.

Principle of Non-Contradiction: This principle states that a proposition cannot be both true and false at the same time in the same context. Two opposing statements cannot be held true simultaneously.

Principle of Excluded Middle: This principle asserts that a proposition must be either true or false; there is no third option. In other words, there is no middle ground between truth and falsity.

Principle of Sufficient Reason: This principle maintains that for every event or proposition, there must be a sufficient reason or cause that explains why it is so and not otherwise. It is used to argue that everything has an explanation.

Rules of Inference: These are logical rules that apply to derive valid conclusions from premises. Examples of rules of inference include modus ponens, modus tollens, and conjunction elimination.

Distributive Law: This law applies to logical operators such as "and" and "or" and establishes how they relate in logical expressions. For example, the distributive law states that A and (B or C) is equivalent to (A and B) or (A and C).

Law of Exclusion: This law states that A or (not A) is always true. In other words, a proposition or its negation is always true.

De Morgan's Laws: These laws apply to the negations of conjunctions and disjunctions. They help simplify complex logical expressions. For example, the negation of (A and B) is equivalent to (not A or not B).

Laws of Implication: These laws govern the relationship between a proposition and its negation, as well as between a proposition and its converse. Examples of implication laws include the law of negation and the law of contraposition.

These rules and principles form the foundation of logic and are used to evaluate the validity of arguments and logical expressions. By following these rules, it is possible to determine whether a reasoning is logically sound and if the conclusions are consistent with the premises. Logic is an essential tool in critical thinking and informed decision-making.

Valid Arguments: A valid argument in logic is one in which, if the premises are true, the conclusion must be true. Logic is used to evaluate the validity of arguments and ensure that conclusions logically follow from the premises. The notion of valid arguments is fundamental in logic. A valid argument is one in which, if all premises are true, then the conclusion must be true. In other words, a valid argument has a logical structure such that if we accept that all premises are true, we cannot help but conclude that the conclusion is also true. This guarantees a strong logical relationship between the premises and the conclusion.

A classic example of a valid argument is modus ponens, which follows this structure: If A is true, then B is true. A is true. With these premises, we can validly conclude: Therefore, B is true.

In this case, if we accept that premise 1 and premise 2 are true, conclusion 3 must be true, according to the rules of logic. This is an example of a valid argument.

However, it is important to highlight that the validity of an argument does not guarantee that the premises are true in reality. Validity refers to the logical relationship between the premises and the conclusion, but the premises themselves must be verified based on evidence and factual truth.

Valid arguments are an essential component of critical thinking and effective argumentation, as they allow for the construction of solid and coherent reasoning. When evaluating arguments in logic, the goal is to determine whether they are valid and, if so, whether the premises are true or plausible. This is crucial in making informed decisions and evaluating claims and arguments in various contexts.

Propositions and Logical Connectives: In logic, we work with propositions, which are assertions or statements that can be true or false. Logical connectives, such as "and," "or," and "if... then," are used to combine propositions and construct arguments. In logic, propositions and logical connectives are fundamental elements used to construct arguments and logical expressions.

Propositions:

Proposition: A proposition is an assertion or statement that can be evaluated as true or false, but not both at the same time. For example, "The sky is blue" is a proposition, as it can be true (during the day) or false (at night).

Validity of Propositions: In logic, propositions are considered basic units of information. They can be simple (a single assertion) or compound (combinations of simple propositions using logical connectives).

Logical Connectives:

"And" (Conjunction): The connector "and" is used to combine two propositions, and the compound proposition is true only if both simple propositions are true. For example, "It is raining and it is cold" is true only if both assertions are true.

"Or" (Disjunction): The connector "or" is used to combine two propositions, and the compound proposition is true if at least one of the simple propositions is true. For example, "I will study math or English" is true if I plan to study at least one of those subjects.

"If... Then..." (Implication): Implication is used to express a conditional relationship between two propositions. The compound proposition is true unless the first proposition is true and the second is false. For example, "If it rains, then I will take an umbrella" is true unless it rains and I do not take an umbrella.

"If and only if" (Biconditional): This connector is used to indicate that two propositions are true or false together. The compound proposition is true if both propositions are the same (both true or both false). For example, "I will succeed if and only if I work hard" is true only if both assertions are true or both are false.

"Not" (Negation): Negation is used to express the opposite of a simple proposition. For example, "It is not true that the sun is blue" negates the assertion that the sun is blue.

These logical connectives are essential for building arguments, logical expressions, and rules in logic and mathematics. They are used to combine propositions and build logical structures that allow reasoning and evaluation of arguments based on solid logical principles. Logic is utilized in a wide range of applications, from mathematics and philosophy to programming and decision-making in the real world.

Deduction and Induction: Logic includes both deductive and inductive reasoning. Deductive reasoning is based on rules to reach necessarily true conclusions, while inductive reasoning is based on probability and generalization from examples. Logic encompasses two fundamental types of reasoning: deductive reasoning and inductive reasoning. Each has its own characteristics and applications:

Deductive Reasoning:

Deductive reasoning is based on logical rules and principles to reach necessarily true conclusions from given premises.

In deductive reasoning, if the premises are true and the structure of the argument is valid, the conclusion must be true. This is known as "necessary truth."

Example: All men are mortal (premise 1), Socrates is a man (premise 2), therefore, Socrates is mortal (conclusion).

Inductive Reasoning:

Inductive reasoning is based on the observation of examples and generalization from them. Conclusions in inductive reasoning are not necessarily true, but they are probable or reasonable based on the available evidence.

In inductive reasoning, patterns or trends are inferred from observed examples and extended to general conclusions.

Example: I observe that every time I flip a coin, it lands heads up. Therefore, I can inductively conclude that the coin is unfair and will always land heads up.

Both types of reasoning are important and have applications in different contexts:

Deductive reasoning is used to establish conclusions based on solid logical rules. It is essential in mathematics, philosophy, and legal argumentation, where logical validity is critical.

Inductive reasoning is used to infer generalizations based on observations and experience. It is often applied in science, social research, and everyday decision-making. However, inductive conclusions always carry some degree of uncertainty due to their probabilistic nature.

Both types of reasoning have their advantages and limitations. Deductive reasoning provides solid and certain conclusions, but it depends on the validity of the premises. Inductive reasoning allows for useful conclusions based on experience, but it can lead to errors if not based on a representative sample or if observations are biased. Both types of reasoning are essential tools in logic and rational decision-making.

Logical Forms: Logical forms are patterns of reasoning that follow specific logical rules. Identifying the logical form of an argument is useful for evaluating its validity regardless of the specific content. Logical forms are abstract patterns of reasoning that follow specific logical rules and allow for the evaluation of an argument's validity independently of its concrete content. Identifying the logical form of an argument is a valuable technique in logic and critical thinking, as it enables the determination of whether the structure of the reasoning is solid and coherent.

Modus Ponens: This logical form follows the structure: If A, then B. A. Therefore, B.

Modus Tollens: Another logical form that follows this structure: If A, then B. Not B. Therefore, not A.

Syllogism: A syllogism is a logical form consisting of three propositions: two premises and a conclusion. There are different types of syllogisms, such as categorical syllogisms or hypothetical syllogisms. An example of a categorical syllogism would be: All people are mortal (Premise 1). Socrates is a person (Premise 2). Therefore, Socrates is mortal (Conclusion).

Inclusive Disjunction: This logical form follows the structure: A or B. It is not the case that A and B are false. Therefore, at least one of the two propositions is true.

Reduction to Absurdity (Modus Tollendo Ponens): A technique in which the negation of the conclusion is assumed and a contradiction is derived, demonstrating that the negation of the conclusion is false. This, in turn, demonstrates that the original conclusion is true.

Conditional: This logical form follows the structure: If A, then B. B is false. Therefore, A is false.

Identifying the logical form of an argument is useful for evaluating its validity and detecting possible fallacies. It is often utilized in formal logic and in the analysis of arguments in philosophy, mathematics, and other fields where the validity of reasoning is essential. The focus on logical form allows for simplifying the evaluation of arguments by separating the logical structure from their concrete content, making it easier to identify valid and invalid arguments.

Syllogisms: Syllogisms are logical arguments consisting of two premises and a conclusion. Logic is used to evaluate the validity of syllogisms. Syllogisms are logical arguments that follow a specific structure composed of two premises and a conclusion. This form of reasoning is used to reach a conclusion based on the relationship between the premises. Syllogisms are an important element in logic and are used to evaluate their validity.

A typical syllogism consists of the following parts:

Major Premise: The first premise in a syllogism, often a general or universal statement.

Minor Premise: The second premise, which provides specific information or a statement about an individual case.

Conclusion: The statement derived from the two premises.

Syllogisms are classified into different types based on the form of their premises and conclusions. Some examples of categories of syllogisms include:

Categorical Syllogism: Uses categorical propositions to form arguments. Example: Major Premise: All men are mortal. Minor Premise: Socrates is a man. Conclusion: Socrates is mortal.

Hypothetical Syllogism: Uses conditional or hypothetical propositions in its premises. Example: Major Premise: If you study, then you will learn. Minor Premise: You are studying. Conclusion: You will learn.

Disjunctive Syllogism: Based on disjunctive (alternative) propositions. Example: Major Premise: The exam will be in math or history. Minor Premise: It will not be in math. Conclusion: The exam will be in history.

Logic is used to evaluate the validity of syllogisms. A valid syllogism is one in which the conclusion necessarily follows from the premises. However, a syllogism can be valid but not necessarily true if any of the premises are false. Therefore, in addition to evaluating validity, it is also important to verify the truth of the premises in the context of the argumentation.

Syllogisms are useful tools in argumentation, philosophy, rhetoric, and decision-making, as they allow for structured reasoning and the evaluation of the strength of arguments.

Application in Mathematics and Philosophy: Logic plays a crucial role in mathematics and philosophy. In mathematics, it is used to prove theorems and validate mathematical propositions. In philosophy, it is used to analyze arguments and concepts. Logic is a fundamental tool in both mathematics and philosophy:

Mathematics:

Mathematical Proofs: In mathematics, logic plays an essential role in formulating rigorous mathematical proofs. Mathematicians use logic to construct logical arguments that demonstrate the truth or falsity of theorems and mathematical propositions. These proofs are based on solid logical principles and coherent argumentative structures.

Set Theory: Logic is also used in set theory, where logical operations, such as union, intersection, and difference, are applied to sets to analyze their relationships and properties.

Boolean Algebra: In computer science and the design of logical circuits, Boolean algebra, which is a branch of logic, is used to manipulate and analyze logical expressions and create systems of Boolean logic.

Philosophy:

Argument Analysis: In philosophy, logic is used to analyze and evaluate arguments. Formal logic is applied to determine whether an argument is valid or invalid, and whether its premises logically support its conclusion.

Philosophy of Logic: The philosophy of logic is a branch of philosophy that focuses on issues related to the nature of logic, its validity, applicability, and limitations. Philosophers of logic explore fundamental questions about the structure of valid arguments and the philosophical implications of logic.

Epistemology: Logic is also relevant in epistemology, which is the branch of philosophy concerned with knowledge and belief. Logic is used to analyze the foundations of knowledge and the justification of beliefs.

In both mathematics and philosophy, logic provides a solid foundation for reasoning and argumentation. It facilitates clarity in thought, the identification of fallacies, and the construction of valid arguments. Logic is an essential tool in the pursuit of truth, problem-solving, and critical analysis in these fields.

Problem Solving: Logic is essential in problem-solving. It is applied to break down problems into more manageable parts, identify relationships, and find coherent solutions. Logic is undoubtedly a fundamental tool in problem-solving across a wide variety of fields and situations.

Problem Decomposition: Logic is used to break down a complex problem into smaller, more manageable parts. This involves analyzing the problem, identifying its essential components, and understanding how they relate to each other.

Relationship Identification: Logic is fundamental for identifying relationships and connections between the elements of the problem. This may involve recognizing patterns, identifying causes and effects, or determining rules and constraints governing the problem.

Option Analysis and Evaluation: Logic is applied to evaluate various options or approaches to address a problem. Logical principles help determine which solutions are the most coherent and effective.

Fallacy Detection: Logic is essential for identifying fallacies or errors in reasoning that may arise during problem-solving. By being aware of fallacies, one can avoid traps in thinking and arrive at more accurate solutions.

Constructing Solid Arguments: In problem-solving, it is often necessary to present solid arguments to support a proposed solution. Logic is used to build strong and persuasive arguments that defend a particular solution.

Decision-Making: Logic is essential in the decision-making process. It helps evaluate available options, weigh the pros and cons, and select the option that best fits the goals and constraints of the problem.

Critical Thinking: Critical thinking, which relies on logic, is a crucial skill in problem-solving. It enables questioning, analyzing, and evaluating information rigorously and objectively.

Optimization: In problem-solving, logic is also applied to find optimal or efficient solutions, meaning those that achieve the best outcome based on certain criteria.

Logic is a tool utilized in most fields of study and disciplines, whether in mathematics, science, engineering, business, law, philosophy, computer science, and many others. It facilitates informed decision-making and the search for effective solutions to complex problems.

Programming and Computing: In programming, logic is used to create algorithms and control structures in decision-making. The logic of programming is fundamental in software development and task automation.

Logic is utilized to solve computational problems. It is used to design efficient algorithms that effectively solve problems.

Control Structures: Control structures in programming, such as conditional statements (if/else) and loops (for, while), are based on logic to make decisions and control the flow of the program. Logic is used to define the conditions under which certain parts of the code are executed.

Logical Programming: In programming languages like Prolog, logical programming is used to solve problems involving logical rules and relationships. These languages allow expressing logical relationships directly in the code.

Problem Solving: Programming involves problem-solving, and logic is essential for analyzing and addressing those problems. Programmers use logic to break down problems, identify solutions, and design algorithms that solve specific issues.

Database Design: In database management, logic is used to design database schemas, define relationships between tables, and establish logical integrity constraints.

Automated Reasoning: In fields such as artificial intelligence and machine learning, logic is used to develop systems that can reason logically and make decisions based on data and logical rules.

Formal Programming Languages: Formal programming languages, such as lambda calculus and first-order logic, are used in programming theory to model and analyze program behaviors and logical properties.

Program Verification: Formal verification is a field that uses logic to mathematically demonstrate the correctness of programs. It is applied to critical systems where errors can have serious consequences, such as aviation or the medical industry.

The logic of programming is essential in software development and in task automation, as it allows programmers to create logically coherent systems and make decisions based on specific conditions. Logical programming is a fundamental component in problem-solving in computing and the creation of high-quality software.

Decision-Making: Logic also plays a role in informed decision-making. It helps evaluate options and determine the most logical or reasonable choice.

Option Evaluation: Decision-making involves considering several options and evaluating them. Logic is used to analyze each option objectively, taking into account its pros and cons. This involves applying logical principles to determine the validity of claims and the strength of arguments supporting each option.

Consequence Identification: Logic is used to foresee the consequences of each option. It helps determine how decisions may affect different variables or outcomes. This logical evaluation of consequences is fundamental for making informed decisions.

Cost-Benefit Analysis: Logic is applied in cost-benefit analysis, where the costs and benefits associated with each option are evaluated. This involves logically comparing costs with benefits to determine which option is the most reasonable from a financial perspective.

Deductive Reasoning: Deductive logic is used to reason from known premises to specific conclusions. This type of reasoning is valuable when making decisions based on available and known information.

Inductive Reasoning: Inductive reasoning, which is based on probability and generalization from examples, is used when decision-making involves uncertainty. It helps estimate probabilities and evaluate different possible scenarios.

Avoiding Fallacies: Logic is also applied to avoid fallacies or errors in reasoning in the decision-making process. Critical thinking, which is based on logic, helps identify thought traps that could influence incorrect decisions.

Argument Formulation: In situations where decisions need to be justified to others, logic is used to construct solid arguments that support the choice made. This is especially important in professional environments and collective decision-making contexts.

Ethics and Values: Logic also plays a role in ethical decision-making. It helps analyze and evaluate ethical arguments, identify value conflicts, and reach decisions consistent with an ethical framework.

Logic is a valuable tool in decision-making, as it allows for a structured and evidence-based approach to evaluate options and determine the best alternative. It helps minimize the influence of biases and emotions in decisions, resulting in more rational and well-founded choices. It is an essential discipline in many areas of knowledge and a tool for problem-solving, effective argumentation, and decision-making based on solid reasoning. Its application is diverse and spans from mathematics to philosophy, computer science, and everyday life.